Wissenschaftsethik und Technikfolgenbeurteilung
Band 11

Schriftenreihe der Europäischen Akademie zur Erforschung
von Folgen wissenschaftlich-technischer Entwicklungen
Bad Neuenahr-Ahrweiler GmbH
herausgegeben von Carl Friedrich Gethmann

Springer-Verlag Berlin Heidelberg GmbH

M. Decker (Ed.)

Interdisciplinarity in Technology Assessment

Implementation and its Chances and Limits

Mit 15 Abbildungen und 10 Tabellen

 Springer

Reihenherausgeber
Professor Dr. Carl Friedrich Gethmann
Europäische Akademie GmbH
Wilhelmstraße 56, 53474 Bad Neuenahr-Ahrweiler, Germany

Bandherausgeber
Dr. Michael Decker
Europäische Akademie GmbH
Wilhelmstraße 56, 53474 Bad Neuenahr-Ahrweiler, Germany

Redaktion
Friederike Wütscher
Europäische Akademie GmbH
Wilhelmstraße 56, 53474 Bad Neuenahr-Ahrweiler, Germany

ISBN 978-3-642-07671-8

Die Deutsche Bibliothek-CIP-Einheitsaufnahme
Interdisciplinary in technology assessment: implementation and its chances and limits /
ed.: Michael Decker.... - Berlin; Heidelberg; New York; Barcelona; Hongkong; London; Mailand;
Paris; Tokio: Springer, 2001
    (Wissenschaftsethik und Technikfolgenbeurteilung; Bd. 11)
    ISBN 978-3-642-07671-8   ISBN 978-3-662-04371-4 (eBook)
    DOI 10.1007/978-3-662-04371-4

http://www.springer.de

© Springer-Verlag Berlin Heidelberg 2001
Originally published by Springer-Verlag Berlin Heidelberg New York in 2001
Softcover reprint of the hardcover 1st edition 2001

Reproduktionsfertige Vorlagen von: Europäische Akademie, Bad Neuenahr
Einbandgestaltung: de'blik, Berlin
SPIN: 10856217   Gedruckt auf säurefreiem Papier   62/3020hu - 5 4 3 2 1 0

Europäische Akademie

zur Erforschung von Folgen wissenschaftlich-technischer Entwicklungen
Bad Neuenahr-Ahrweiler GmbH

Direktor
Professor Dr Carl Friedrich Gethmann

The Europäische Akademie

The *Europäische Akademie GmbH* is concerned with the scientific study of consequences of scientific and technological advance for the individual and social life and for the natural environment. The Europäische Akademie intends to contribute to a rational way of society of dealing with the consequences of scientific and technological developments. This aim is mainly realised in the development of recommendations for options to act, from the point of view of long-term societal acceptance. The work of the Europäische Akademie mostly takes place in temporary interdisciplinary project groups, whose members are recognised scientists from European universities. Overarching issues, e. g. from the fields of Technology Assessment or Ethics of Science, are dealt with by the staff of the Europäische Akademie.

The Series

The series "Wissenschaftsethik und Technikfolgenbeurteilung" ("Ethics of Science and Technology Assessment") serves to publish the results of the work of the Europäische Akademie. It is published by the Academy's director. Besides the final results of the project groups the series includes volumes on general questions of ethics of science and technology assessment as well as other monographic studies.

Acknowledgement

This publication was supported by the European Commission, Research DG, Human Potential Programme, High-Level Scientific Conferences (HPCF-CT-2000-00279). The information provided is the sole responsibility of the authors and editor and does not reflect the Community's opinion. The Community is not responsible for any use that might be made of data appearing in this publication.

Foreword

The Europäische Akademie zur Erforschung von Folgen wissenschaftlich-technischer Entwicklungen Bad Neuenahr-Ahrweiler GmbH (European Academy for the study of consequences of scientific and technological advance) is concerned with the scientific study of the consequences of scientific and technological advance for the individual and social life and for the natural environment. The main focus is to examine foreseeable mid- and long-term processes that are especially influenced by natural- and engineering sciences and the medical disciplines. The academy fulfills this task by organizing interdisciplinary expert discussions.

Another important issue of the work of the Europäische Akademie concerns the methodology of Technology Assessment as a general issue. It is mainly realized by the academy's staff. In this connection the academy strives for a broader discussion with external experts; these discussions are held at the scientific conferences which take place twice a year.

Following the spring and autum conferences in 1997, "Technology Assessment and Policy Consulting" and "Ethics in Technical Acting. Practical Relevance and Legitimation", the autumn conference "Implementations and Limits of Interdisciplinarity in European Technology Assessment" (September 2000 at Bad Neuenahr-Ahrweiler) was the third meeting dealing with methodological aspects of Technology Assessment.

At this conference scientists from sixteen European Countries discussed the topic of interdisciplinarity which is the core notion of the Europäische Akademie's working method. They considered aspects like quality control, the relevance for policy consulting, the participation of laypersons, stakeholders and/or non-governmental organizations. Thanks to the financial support by the European Commission young researchers could be invited to attend the conference.

The book series on ethics of science and technology assessment, "Wissenschaftsehtik und Technikfolgenbeurteilung", edited by the Europäische Akademie, gives the opportunity to publish the results of this conference as a contribution to the ongoing debate on the methodology in European Technology Assessment.

Bad Neuenahr-Ahrweiler, July 2001 *Carl Friedrich Gethmann*

Preface

Interdisciplinarity and Technology Assessment (TA) are widely accepted as inseparably connected. This is due to the fact that TA has been established to solve social, political or ecological problems and, in general, these problems are not solvable by an individual scientific discipline alone. However, a more precise view produces a less homogeneous idea of interdisciplinarity in TA. Concerning operationalisation of interdisciplinarity, there is even general confusion on this issue.

The contributions in this book originate from the conference "Implementations and Limits of Interdisciplinarity in European Technology Assessment", organised by the Europäische Akademie in September 21–23, 2000 at Bad Neuenahr-Ahrweiler. The conference was organised as an initialising discussion forum. This was realised by both a prepared comment on every key note and by sufficient time for discussion in the plenum. The keynote lecturers had to present their main arguments to the commentators in advance. After the conference all authors were asked to consider the main arguments of the discussion in their contributions. Therefore, this book claims to be more than the "proceedings of a conference". Every topic is treated either in the form of a presentation followed by critical response or from two perspectives. In this way the reader gets an insight in the ongoing discussion.

All authors took on this unusually strenuous procedure of intensive discussion and submitted their contributions punctually, special thanks for this effort. Fruitful discussion needs experienced chair persons. This part was commendably taken over by Dr. Sergio Bellucci (Centre for TA at the Swiss science and technology council), by Dr. Henk van de Graaf (University of Amsterdam) and by Dr. Walter Peissl (Institute of Technology Assessment (ITA) of the Austrian Academy of Sciences).

Special thanks is due to Professor Dr. Armin Grunwald from the Institute for Technology Assessment and System Analysis (ITAS) in Karlsruhe and to Dr. Felix Thiele from the Europäische Akademie, who were, together with the editor of this volume, members of the scientific committee and who participated in planning the conference and in selecting the lecturers. Their successful projection convinced the European Commission to support the conference as "High Level Conference" in their Human Potential Programme.

Bad Neuenahr-Ahrweiler, June 2001   *Michael Decker*

List of Authors

Carrier, Martin, Dr., studied physics and philosophy at the Universität Münster. Ph. D. 1984 in philosophy at the same university on a thesis on methodological features of 18th century affinity theory. Research on theory-ladenness of data and space-time philosophy at the Universität Kontanz. Habilitation 1989 with a thesis on the relation between theory and evidence in space-time theories (published 1994 as The Completeness of Scientific Theories). Between 1994 and 1998 professor for philosophy of science at the Universität Heidelberg. Since 1998 in the same position at the Universität Bielefeld. Main areas of research: theory change, intertheoretical relations.
Postal address: Department of philosophy, Universität Bielefeld, P.O.B. 100 131, 33501 Bielefeld, Germany

Cope, David, MA, Cambridge University, 1967, MSc (Econ.), London School of Economics, 1968, Lecturer in Interdisciplinary Studies, Nottingham University, 1970–81; Environmental Team Leader, International Energy Agency of the OECD, 1981–86; Executive Director, UK Centre for Economic and Environmental Development, Cambridge, England, 1986–1997; Professor of Energy Economics, Doshisha University, Kyoto, Japan, 1997–98; Director, Parliamentary Office of Science and Technology, Houses of Parliament, London, England, 1998–present. Main research interests: energy and environment policy, especially electricity generation; Technology Assessment; economy and society of North-East Asia.
Postal address: Parliamentary Office of Science and Technology, Westminster House, House of Commons, 7 Millbank, Westminster, London SW1P 3JA, United Kingdom

Decker, Michael, Dr., studied physics (minor subject economics) at the Universität Heidelberg, 1992 diploma, 1995 doctorate with a dissertation on temperature measurements in high pressure combustion by laser-induced fluorescence of molecular oxygen at the Universität Heidelberg, 1995–1997 scientist at the German Aerospace Centre (DLR) in Stuttgart, since February 1997 member of the scientific staff of the Europäische Akademie, project coordinator of the project group "Robotics. Options of the replaceability of human beings" and member of the study group "practical philosophy". Main research

areas: Methodology of Technology Assessment, Concepts of Interdisciplinarity.
Postal address: Europäische Akademie zur Erforschung von Folgen wissenschaftlich-technischer Entwicklungen Bad Neuenahr-Ahrweiler GmbH, Wilhelmstr. 56, D-53474 Bad Neuenahr-Ahrweiler, Germany

Funtowicz, Silvio taught mathematics, logic and research methodology in Buenos Aires, Argentina. During the decade of 1980 he was a Research Fellow at the University of Leeds, England. He is now a Scientific Officer at the European Commission Joint Research Centre in Ispra, Italy, where he is in charge of the programme of Knowledge Assessment at the Institute of Systems, Informatics and Safety. He is the author of Uncertainty and Quality in Science for Policy (1990, Kluwer, Dordrecht) in collaboration with Jerry Ravetz, and numerous papers in the field of environmental and technological risks and policy-related research. He is a member of the editorial board of several publications and of the scientific committee of many international conferences, and has lectured extensively.
Postal address: European Commission – Joint Research Centre – Institute for Systems, Informatics and Safety, TP 650, 21020 Ispra (VA), Italy

Gethmann, Carl Friedrich, Professor Dr., studies of philosophy at the universities of Bonn, Innsbruck and Bochum; 1974 Dissertation on methodology in Martin Heidegger at Universität Bochum; 1979 Habilitation on formal pragmatics of substantiatory discourses at Universität Konstanz; 1968 Scientific assistant; 1972 Lecturer for philosophy at the Universität Essen; further lectureships at the universities of Düsseldorf and Göttingen. – Appointment to a full professorship (C 4) at the Universität Oldenburg (1990), at the Academy of Technology Assessment in Stuttgart (1991) as well as at the universities of Essen (1991), Konstanz (1993) and Bonn (1995); Full professorship for philosophy at the Universität Essen (1991); Director of the Europäische Akademie zur Erforschung von Folgen wissenschaftlich-technischer Entwicklungen Bad Neuenahr-Ahrweiler GmbH (1996); Member of the Academia Europaea (London); full member of the Berlin-Brandenburgische Akademie der Wissenschaften. Main research areas: Philosophy of language/ Philosophy of logic, Phenomenology, Applied philosophy/Technology Assessment.
Postal address: Europäische Akademie zur Erforschung von Folgen wissenschaftlich-technischer Entwicklungen Bad Neuenahr-Ahrweiler GmbH, Wilhelmstr. 56, D-53474 Bad Neuenahr-Ahrweiler, Germany

Grunwald, Armin, Professor Dr., studied physics at the universities of Münster and Cologne, 1984 diploma, 1987 dissertation on thermal transport processes in semiconductors at Universität Köln, 1987–1991 systems specialist, studies of mathematics and philosophy at Universität Köln, 1992 graduate (Staatsexamen), 1991–1995 scientist at the DLR (German Aerospace Cen-

ter) in the field of Technology Assessment, since 1996 vice director of the Europäische Akademie, 1998 habilitation at the faculty of social sciences and philosophy at Universität Marburg with a study on culturalistic planning theory. Since October 1999 director of the institute for technology assessment and systems analysis (ITAS) at the research center Karlsruhe and professor for Technology Assessment at the Universität Freiburg.
Postal address: Institut für Technikfolgenabschätzung und Systemanalyse, Forschungszentrum Karlsruhe, Postfach 36 40, D-76021 Karlsruhe, Germany

Hronszky, Imre, Professor Dr., MSc in chemistry, University Eotvos/Budapest 1965, MSc (Phil.), University Eotvos/Budapest 1969, Dr. rer. nat., 1973, PhD in philosophy 1983, Dr. Habil. in engineering sciences, 1998, recently professor at the Budapest University of Technology, founder and head of department of Innovation Studies and History of Technology. Publications in five languages in philosophy of science, history of chemistry, Technology Assessment, innovation studies, methodology of history of technology. Main recent research interests: socially acceptable innovation policy, basic problems of Technology Assessment, the role of history of technology for innovation studies.
Postal address: BUTE, 1111 Sztoczek u.2, Hungary

Jovell, Albert J. holds an MD and a PhD in Sociology degrees from the University of Barcelona (Spain); a Master of Public Health (MPH) in Epidemiology, a Master of Science (MS) in Health Policy and Management, and a Doctor of Public Health (DPH) degrees from Harvard University (USA), and a MA degree in Political and Social Sciences from the University Autonoma of Barcelona (Spain). He is the Director General of the Josep Laporte's Library Foundation and Associate Professor of the Department of Preventive Medicine and Public Health of the School of Medicine of the University Autonoma of Barcelona. He was head of the Research and Academic Unit of the Catalan Agency for Health Technology Assessment (CATHA) and worked at the pharmaceutical industry and in the Technology Assessment Group at Harvard School of Public Health. He served as Board of Directors of the International Society for Technology Assessment in Health Care (1997–2000).
Postal address: Fundació Biblioteca Josep Laporte, Sant Antoni Ma Claret, 171, 08041 Barcelona, Spain

Kemp, René, PhD, is senior research fellow at MERIT in Maastricht and research director of STEP in Oslo. Trained as an econometrician, he became a policy analyst and innovation researcher. He is interested in empirical analysis and theorizing, with a special interest in macro-governance issues. René Kemp has published widely in the area of economics of technical change and environmental economics and his PhD thesis on environmental policy and technical change, published by Edward Elgar in 1997, is viewed a major contribution on the topic. His research interests are: environmental policy and technical change, technological transitions: the institutional aspects, technological regime shifts to environmental sustainability, green innovation policy and evolutionary theories of technical change.

Postal address: Maastricht Economic Research Institute on Innovation and Technology, Maastricht University, P.O. Box 616, 6200 MD Maastricht, The Netherlands

Kloprogge, Penny, Ir., MSc degree in environmental sciences at Wageningen Agricultural University, The Netherlands, 1998. Since 1999 PhD-student at Utrecht University, Dept. of Science, Technology and Society. PhD research on coping with value diversity in Integrated Assessment modelling of Climate Change.

Postal address: Department of Science Technology and Society, Utrecht University, Padualaan 14, 3584 CH Utrecht, The Netherlands

Liakopoulos, Miltos, PhD, studied Psychology in Athens (BA) and Social Psychology in London (MSC, PhD) with a thesis on the analysis of the debate on biotechnology; since 1994 he has held various positions as researcher at the London School of Economics and the Science Museum, London, in the area of public perceptions of Science & Technology in general and biotechnology in particular; since 2000 he is member of the scientific staff at the Europäische Akademie co-ordinating the project "Functional Foods"; his research interests are in the areas of science & technology perceptions and policy, impact of TA in policy making, and ethics of science.

Postal address: Europäische Akademie zur Erforschung von Folgen wissenschaftlich-technischer Entwicklungen Bad Neuenahr-Ahrweiler GmbH, Wilhelmstr. 56, D-53474 Bad Neuenahr-Ahrweiler, Germany

Ravetz, Jerry is a leading authority on the history and philosophy of science in the policy domain. His seminal work *Scientific Knowledge and its Social Problems* (Oxford 1971) was republished by Transaction Publishers in 1996. He has worked with Silvio Funtowicz in creating the NUSAP system for the management of uncertainty, and on Post-Normal Science. He is currently Director of the Research Methods Consultancy of London, and the convenor of the UK Governance & Science Group.

Postal address: RMC Ltd, 196 Clarence Gate Gardens, London, United Kingdom

Renn, Ortwin, Professor Dr., studied Sociology, Economics, Social Psychology and Journalism in Cologne. He serves as full professor at the Universität Stuttgart and chairs the department of Environmental Sociology. He is also Chair of the Board of Directors at the Center of Technology Assessment in Baden-Württemberg, in which he directs one of the four Center's departments entitled: Technology, Society and Environmental Economics. Dr. Renn chairs

Together with Jan Rotmans she developed an interdisciplinary method to deal with uncertainty in IA modelling. From 1996 till 1997 she worked at the Swiss Institute for Environment and Technology (EAWAG). Marjolein van Asselt is co-founder of the International Centre for Integrative Studies (ICIS) founded in November 1997 and of ICIS BV. In 2000 she obtained her PhD at Maastricht University, the Netherlands, in 'Perspectives on Uncertainty and Risk'. She is currently working as senior researcher IA methodology. She is a member of the Dutch Scientific Council on Environment Nature and Spatial Research (RMNO).
Postal address: International Centre for Integrative Studies, P.O. Box 616, 6200 MD Maastricht, The Netherlands

van der Sluijs, Jeroen P., PhD, MSc degree in theoretical ecology, Leiden University 1990, PhD 1997 at Utrecht University on a thesis on the management of uncertainties in risk assessment of anthropogenic climate change. Since then assistant professor and co-ordinator of the research group "Risk Assessment and Risk Management" within the Department of Science Technology and Society, Utrecht University. Main research areas: management of uncertainty and value-ladenness in science for policy, participatory approaches to risk assessment, Integrated Assessment, climate change policies, burden differentiation.
Postal address: Department of Science Technology and Society, Utrecht University, Padualaan 14, 3584 CH Utrecht, The Netherlands

Contents

II Technology Assessment as Policy Consulting   71

I Methodology of Technology Assessment

1 Participatory Technology Assessment. Some Critical Questions

Carl Friedrich Gethmann

The central theme of this conference is devoted to the question of the implementation and limits of inter-disciplinary co-operation in TA. To many of those present, this question may certainly appear to be somewhat out-dated. Since, after all, it presupposes that TA is connected, or at least that essential parts of it are connected, with (scientific) disciplines. By this it is presupposed that TA as a whole, or at least in its essential parts, is a scientific undertaking. This is the very thing that is disputed by so called "modern" concepts. These concepts, which can be summarised under the heading of "Participatory TA", profess to have succeeded in overthrowing the elitist, expertocratic claim of scientific TA in favour of a democratic, communicative form involving the participation of the citizens.[1] Thus, systematically, before the question of interdisciplinary co-operation can be answered, the question has to be addressed as to whether TA is primarily a scientific or an extra-scientific (societal, communicative) undertaking. The question of interdisciplinary co-operation is fundamentally dependent on the answer to this preliminary question. This all the more so since the impression exists in the meantime in a number of European countries, e. g. the Netherlands and Denmark, that the participatory element constitutes the paradigm of TA as a whole.[2] Thus it might well be appropriate to put this question explicitly at the beginning of this conference. I shall address the issue by posing some critical questions concerning those concepts of participatory TA. These will allude to the concept of scientific TA. Anyone who follows this trend, however, must subsequently be ready for the question as to the manner in which the scientific status of TA shall be determined. Hence the topic of interdisciplinary co-operation is now a topic for debate.

As it is widely known, the discussion with regard to the various concepts of TA is not exactly characterized by a clear classification of these concepts. For this reason, the concepts of TA are mostly depicted following a chrono-

---

[1] For an overview cf. Gethmann and Grunwald (1996); Grunwald (1999); Köberle et al. (1997); Gottschalk and Elstner (1997).

[2] Eijndhoven and Est (1999) is an example for the Netherlands, Klüver (1995) and Meyer (1999) are more or less typical representants for Denmark.

logical rather than a systematic enumeration.[3] In particular, a great deal of terminological effort would be needed to differentiate between participatory, public, constructive, interactive, and discursive TA. I shall dispense with this differentiation and attempt at once to formulate a fundamental concept for all these different approaches. The principle of participation is obviously directed against concepts which do not envisage any such participation. The participants are, according to the concepts, lay persons, citizens, directly concerned persons, consumers, stake-holders, etc. on the one hand, as compared with (technical) experts, scientists, institutional decision-makers, producers, share-holders, etc. on the other. Fundamentally, we are currently experiencing a thrust in the direction of allotting certain areas of decision-making competence (e. g. to the citizens) and withdrawing certain areas of competence (e. g. from the experts). According to the concepts under debate in this process, cognitive, more precisely scientific competence is certainly involved, too. For this reason not simply plebiscitary procedures alone but rather a multitudinous variety of combined procedures are put forward. This is why the many different variations of participatory TA do not merely presuppose a shift of competence (e. g. from scientist to lay person); but rather something like a competence schism along the dividing line between cognitive versus evaluative competence.

Thus, it is ultimately a question of allocating competence.

In the allocation of competence a number of premises and pre-suppositions reveal themselves. All in all, these premises and pre-suppositions lead to the creation of basic, more or less democratic decision-making procedures, or at least of symbols (like consensus conferences), which are in some way reflective of basic democratic decision-making processes. These premises and pre-suppositions can be developed more precisely within three theses:

Thesis I	The sciences are not endowed with any primarily evaluative competence. This premise pre-supposes scientific descriptivism and moral non-cognitivism (*scientific premise*).
Thesis II	The citizen, in contrast to the scientist, is primarily endowed with this evaluative competence (*ethical premise*).
Thesis III	The democratic institutions, constituted by delegation and representation of competence, are unsuitable (at any rate in a number of important cases) to execute the will of the citizens (*political premise*).

[3] 'Participation' resp. 'participatory' are omnipresent catchwords in the debate, often used to characterise different types of TA (Baron 1995; Hennen 1999), amongst them the above mentioned: "Public" TA (Bütschi 2000; Eijndhoven and Est 2000); "Constructive" TA (Schot 1992; Rip et al. 1995), "Interactive" TA (Grin et al. 1997); "Discursive" TA (Daele and Döbert 1995; Döbert 1997).

In the following I shall address these three premises critically, but in order to avoid misunderstanding, before I turn to the theses themselves, which are certainly controversial, I should first like to make it clear which hypotheses I do *not* wish to deny:

a) I do not wish to dispute that in a democratic society all power and thus all decisions on technology policy must ultimately stem from the people (and not from technical, scientific or economic elite).

b) I do not dispute that it is right that the citizens should have a part in community decision-making processes involving consequences for the people concerned. This is especially valid for political decisions which *immediately* affect the citizens themselves.

c) And finally, I do not dispute that it is desirable that the people, on the basis of a good scientific education, are able to understand scientific and technological processes.

1.1
The Tribalisation of Science

First I should like to turn to examining the premise that the sciences are not primarily endowed with evaluative competence. In accordance with this they are then not primarily responsible when it is a matter of technology and scientific policy decision-making on questions with considerable bearing on the people at large or at least on major parts of the population. In my introduction in this connection I already characterised this view as scientific descriptivism and moral non-cognitivism. The questions to be asked here may be discussed from two different points of view:

I. Is science really exclusively or primarily, a purely descriptive, explanatory undertaking?

II. Do scientists, as such, really possess no evaluative competence?

The *first* question leads into the field of general philosophy of science and is widely discussed there. I am unable to expound upon this discussion here.[4] Therefore just a few brief observations: Contrary to the dictum of Max Weber, there are genuinely normative sciences, e.g. economics, jurisprudence and ethics, to name but a few. In addition, the so-called descriptive sciences are riddled with norms of a methodological nature, which in turn must be justified with regard to their rational expedience. These norms belong to the field of competence of the scientists (and not lay persons).

The *second* question first involves conceding that it is not possible to conclude directly from the methodological status of a given science the competence of those engaged in that science. It is *not* valid to assume a priori that

---

[4] For a more detailed discussion cf. Janich et al. (1974).

economists behave more economically, lawyers more lawfully and moral scientists more morally than other people. Although it *is* valid that economists *know* better what is economically rational, jurists *know* better what is in conformity with the law and moral scientists *know* better what is moral. If this is applicable, then it is not permissible to place the scientists (experts, ...) from the point of view of competence as a group alongside other groups (groups of citizens, ...) and to regard them as equal with respect to their competence[5], as an ethnologist would do in placing tribes alongside other tribes, or as a theologist does in setting confessions alongside other confessions. Such a strategy of pluralistic juxtaposition is what I call the "tribalisation" or "confessionalisation" of scientific competence.

The tribalisation of science reveals itself, for example, in the following phenomenon. When scientists deem a toxic risk low and a citizens' action group high, many sociologists interpret it as tantamount to as if a primitive tribe thought rain could be brought about by sacrificing hens, while another tribe were to assert that it must be geese, as if a religious confession believed that there were three sacraments, and another confession were to postulate that there were seven of them. In general: Scientists are seen from without as a social group among others, without special competence being ascribed to them, overlooking the fact that the very inherent defining attribute of the sciences is exactly the self-imposed duty to uphold certain standards of rational reason, which under certain (albeit non-trivial) circumstances permits the claim that a particular assertion of the scientists is true and a demand correct.

Many of the approaches of participatory TA can be seen from my point of view as having fallen victim to the virus of relativism in one of its particularly difficult variants, i. e. that of tribalism. In contrast to this I continue to hold that it is right that though scien*tists* are only human like everyone else, the scien*ces*, on the other hand, provided that certain conditions are fulfilled, can rightly claim acceptability and that the scientists are the ones who are in a position to more or less know best.[6]

---

[5] Cf: "Participative TA is characterised by a procedure in which, in addition to the scientific disciplines, non-scientists, i. e. decision-makers and those directly or indirectly affected by decisions are deliberately involved in the analysis and evaluation process, and are permitted to influence the investigation process, the determination of options for action and/or also the resolving of conflicts by finding solutions through negotiation." W. Baron (1997) p 148 [my translation].

[6] Above all is such a relativism connected with many approaches of sociology of science and science of science. cf. Gethmann (1981).

1.2
The Overtaxing of the Citizens' Competence

In my introduction, the thesis characterised as "ethical premise" contends that the citizen, in contrast to the scientist, is endowed with the evaluative prescriptive competence which is necessary as a basis for making the "right" decisions.[7]

First of all, critical attention must be drawn to an inconsistency in the resulting TA strategies, which can be traced back to unclear interpretations of the word "discourse". If it is primarily a matter of evaluative, prescriptive competence, then why all the effort of "discourse", in which the citizens must be taught the basics of energy technology, molecular biology or reproductive medicine? Programmes in the USA and other countries including Germany which are combined under the heading "education" reflect the misunderstanding existing here.[8] The citizens do not doubt that the biologists are masters of their field and know their business, on the contrary, they are all too afraid that they do; neither do they wish to evaluate the cognitive achievements and performance of the scientists but rather the latter's evaluative prescriptive competence, their reliability and trustworthiness and other character attributes. Trust also has something to do with familiarity with the given topic field, but not with quasi-scientific training. What is to be understood here by "discourse" is urgently in need of precise definition.[9]

But now to the core question: Is the citizen himself endowed with this competence? Is it really appropriate to presuppose that everyone really knows "what is best for himself"? Or isn't it more a matter of postulating that I may expect others to recognise that this is applicable in my particular case? The hypothesis that each individual in regard to his own needs is the best expert to decide what is good for himself, I shall call the "self-competence thesis" ("Eigenkompetenzthese").

Contrary to this thesis stand, first of all, some primary phenomena of life experience, experience gained through the nature of life in its various phases and biological changes from the needs of the child for upbringing and education to the needs of old people for care and attention. In addition, experience is gained from the manipulation and the strategic creation of needs.

Theoretical analysis and interpretation of such experience offers a number of anthropological concepts concerning the factual immaturity of the individual and his inability to determine his own existence. I should like to mention

---

[7] So for example Hennen (1999) p 569: "Participation in TA ... is implicitly committed to a republican model of democracy in which the citizens *themselves are the ones who should make the decisions on those questions which concern them*" [my translation]. In a similar way Kass (2000) p 20 understands participation as a "direct involvement of citizens in political decision-making ... ".

[8] E. g. Akademie für Technikfolgenabschätzung (1995).

[9] Cf. Gethmann and Sander (1999) p 124ff.

Kant's demand for enlightenment as a way out of self-caused disability[10], Marx's concept of false consciousness[11], the concepts of Freud and other psychoanalysts concerning the repression of needs, those theories which have to do with the character of goods as merchandise (e. g. Adorno[12]) and the power-impregnated nature of all discourses (e. g. Foucault[13]).

Naturally, I am not setting out to fundamentally dispute the thesis of self-competence would be justified, not only for the reason that this would mean strategies in whatever political form to place the individual under disability, withdrawing his rights to self-determination. The thesis of self-competence can a priori neither be founded nor refuted. Rather it is a "regulative idea" (in the sense of Kant), which a community must presuppose to be fulfilled, which, however, in the individual case is always only more or less fulfilled. That the citizens, lay persons, persons concerned, ... are endowed with self-competence is therefore not a fact but rather a regulative concept, which must guide us in the organisation of the community but which we must not regard as factually realised.

On the other hand the thesis of factual self-competence constitutes an overtaxing of the evaluative competence of the citizen, from which in many cases he or she rightly would like to see himself or herself freed by the political organisation of the community. By means of delegation and representation procedures, he elects those fellow citizens to whom he wishes to hand over his competence in order to avoid being overburdened by the obligation to exercise his competence in all the decision-making processes. Thus, the idea of grass roots democracy is not at all in the interests of the citizens, who, in fact, for the most part are seeking to fend off the burdens of individual responsibility, but rather in the interest of those political actors who simply want materially different decisions or even a different political system.

1.3
The Plebiscitarism of the Will of the People

The thesis described in my introduction as the political premise contends that the democratic institutions, constituted by delegation and representation of competence, are (in part) unsuitable to execute the will of the people in technical and scientific policy issues.

A particularly conspicuous example is given by the following passage of W. Baron:

> Within a normatively oriented, theoretically democratic framework, participation is designed to afford possibilities for actively taking part in and influencing

---

[10] Kant (1912a) p 33.

[11] The topic of "false consciousness" or "ideology" is widespread over the works of Marx and Engels. For a speaking example cf. Marx and Engels (1969).

[12] Adorno (1980); Horkheimer and Adorno (1988).

[13] Foucault (1971).

political policy-making over and above the act of voting in the elections of a representative democracy, which is increasingly being experienced as insufficient. ... participation constitutes the central principle of action of modern societies, i. e. without opportunities to participate, society would no longer be capable of consensus and thus be incapable of survival.[14]

My initial critical response to such statements, which I should like to make before proceeding any further, is that the denunciatory undertone against representative democracy ("no longer capable of survival") is unignorable. Strategically it is plain to assume that those holding the mandate to represent the people, i. e. the parliamentarians, are not going to particularly promote such demands for self-disempowerment. Utterances of this nature place a considerable burden upon the willingness of the political institutions to embrace the undertaking known as TA. These are, however, as mentioned, merely strategic observations.

As far as the issue itself is concerned, the observation can be made, that the procedures of grass-roots democracy imply a substantial dismantling of the political institutions, above all of the parliament. Since this concept presupposes that, to a certain extent, it is possible to rescind procedures of delegation and representation and the accompanying processes of professionalised decision-making.

At this point I should like to draw attention to a phenomenon which I should like to describe as a "competence allocation cycle". This cycle can be portrayed in a highly simplified stylisation as follows.

- Citizens elect parliaments
- Parliaments call upon the advice of experts
- Experts demand citizens' decisions
- Citizens elect parliaments
- ...

It is plain to see that the substantial dismantling of political representation will result in the substantial dismantling of the very scientific expertise which is needed by the political institutions.

A democratic society is one in which the procedures for forming political will are established in such a way that the filtered will of the people may find its path.[15] Here everything depends on a closer scrutiny of this filter function. The expression "filtered will" is in contrast to "immediate will" of the people. In the long and, indeed, in part bloody history of the development

---

[14] Baron (1997) p 147 [my translation].

[15] This filtering is an effect of the principle of representation that, following Kant, is an essential of each republican form of government (Metaphysik der Sitten (1907) p 339–342; Zum ewigen Frieden (1912b) p 349–353 (Erster Definitivartikel). Within theory of constitutional law the task of these representative processes is called the "increasing and improvement" of the people's will (cf. Krüger (1966) p 232–253, esp. p 233).

of political institutions, European societies have learned that it is not the immediate will of the people which is a guarantor against despotism. Such filter functions are, for example, the division of power, by which means the organs of state power exercise reciprocal control over one another, subsidiarity among the various levels of regional corporations (local, state and federal government bodies) or "time dilatation", i. e. delaying procedures, such as hearings, several readings, etc. So these are the elements which, in the European democracies, belong to the essential structure of the institutions albeit in a wide variety of forms. The structure of democratic institutions in nonplebiscitary democracies expresses, on the one hand, that the power is held by no-one else but the people. On the other hand, the *immediate* will of the people warrants suspicion. Therefore, the institutions must be "forged" in such a way that the people are protected against their own will by means of institutional filters.

But are not these considerations in contradiction to the immediate evidence contained in the demand for participation in cases involving far-reaching decisions, such as the building of supra-regional motor ways or the erection of large-scale plant? I am in no way attempting to dispute that the participation of those directly affected is an essential instrument for improving rational planning in the case of planning procedures for large-scale technical plant. The participation of the citizens in such cases improves communication and by doing so (albeit without being certain) might possibly increase public acceptance, too. But is the legitimisation of the decision itself also improved? This question may also be answered in the affirmative provided that those involved are really directly concerned and affected. But what about the many cases where we have to consider technical and scientific policy decisions in which the degree of being affected is not immediate, where those involved are only indirectly concerned. Here it is hardly possible to draw the line between those genuinely directly concerned and the self-appointed "professional representatives" of the persons concerned ("Berufsbetroffene"). In particular there is no deductive connection between the communicative aspect of participation and the aspect of legitimisation.

Therefore, in conclusion, I should like to express the surmise that the high degree of plausibility which the idea of participation seems to enjoy is essentially thanks to the partially successful employment of participation strategies in citizens' action activities within the framework of planning procedures for large-scale plant. In this sphere, to a certain extent, there occurs a *plausibility transfer* to such areas in which, in principle, everyone is concerned.

If this supposition is correct, then it is decisive to focus more accurately on those sectors in which the idea of participation, especially in consideration of its communicative function, is appropriate, while excluding those issues in which, in any case, the idea of participation has no legitimisation effect. In my opinion, the latter questions include, above all, the highly complex problems accompanying the modern life sciences and the medically-related natural sciences.

Whether it is morally permissible, for example, to clone embryos in order to obtain stem cells for the therapy of a number of serious, hitherto incurable diseases, cannot be, in my view, a problem in which participation should be involved. On the one hand it is the embryos that are directly affected and possibly the donors of the source materials, on the other hand the patients, who are potentially affected.[16] How is one supposed to imagine participation in this case ? What communicative effects can be achieved? What legitimises those who believe that they should raise their voice on this issue?

1.4
Final Remarks

My critical questions have been posed with the aim of focussing on the problem contained in the thesis that TA is primarily a political instrument and not a scientific undertaking. The thesis concerning the political character of TA essentially owes its conception to argumentational premises connected with the concept of participation. Should my doubts with regard to these political notions be justified, one would have to fall back on the old thesis according to which TA is primarily a scientific undertaking.

"Science" here does include the scientific treatment of normative questions as they are objects for disciplines like e. g. philosophical ethics. Presumably the approach of a participative TA could only evolve on the ground of a normative deficit of the approaches in common before. TA which omits normative questions is not feasible indeed.[17]

Admittedly, one must concede that this thesis has lost its plausibility not least because the sciences in their actual state were not in a position to substantiate it. The question as to how science must organise itself in order to make a contribution to technical and scientific policy decisions with their far-reaching consequences has hitherto remained unanswered. A keyword in this connection is interdisciplinary co-operation.

This brings us back to the issue of the realisability and the limits of interdisciplinary co-operation in TA, which is our question for discussion at this conference.

References

Adorno TW (1980)   Minima Moralia. Reflexionen aus dem beschädigten Leben [=Gesammelte Schriften Vol 4]. Frankfurt a. M.

---

[16] Singer et al. (1993); Lauritzen (2000).
[17] Gethmann (1999).

Akademie für Technikfolgenabschätzung in Baden-Württemberg (ed) (1995) Bürgergutachten Biotechnologie/Gentechnik – eine Chance für die Zukunft? Stuttgart

Baron W (1995) Technikfolgenabschätzung. Ansätze zur Institutionalisierung und Chancen der Partizipation. Opladen

Baron W (1997) Grundfragen und Herausforderungen an eine partizipative TA. In: Westphalen R (1997) Technikfolgenabschätzung als politische Aufgabe. München, p 137–158

Bütschi D (2000) The Integration of Lay Expertise in Technology Assessment: Swiss PubliForums as an Example. In: Scholz R, Häberli R, Bill A, Welti M (eds) Transdisciplinarity: Joint Problem-Solving among Science, Technology and Society. Zürich

Daele W van den, Döbert R (1995) Veränderung der äußeren Natur – Partizipative Technikfolgenabschätzung [Funkkolleg Technik, Studienbrief 4, Studieneinheit 11]. Tübingen

Döbert R (1997) Rationalitätsdimensionen von partizipativer Technikfolgenabschätzung. In: Köberle et al. (1997), p 200–213

Eijndhoven J van, Est R van (2000) The Choice of Participatory TA Methods. In: Klüver et al. EUROPTA. Participatory Methods in TA and Technology Decision-Making. Copenhagen.

Est R van, Eijndhoven J van (1999) Parliamentary Technology Assessment at the Rathenau Institute. In: Bröchler S, Simonis G, Sundermann (eds) Handbuch Technikfolgenabschätzung. Berlin, p 427–435

Foulcault M (1971) L'ordre du discours, Paris

Gethmann CF (1981) Wissenschaftsforschung? Zur philosophischen Kritik der nachkuhnschen Reflexionswissenschaften In: Janich P (ed) Wissenschaftstheorie und Wissenschaftsforschung. München, p 9–38

Gethmann CF, Grunwald A (1996) Technikfolgenabschätzung. Konzeptionen im Überblick. (Graue Reihe Nr. 1) Europäische Akademie, Bad Neuenahr-Ahrweiler

Gethmann CF (1999) Die Rolle der Ethik in der Technikfolgenabschätzung In: Petermann T, Coenen R (eds) Technikfolgen-Abschätzung in Deutschland. Bilanz und Perspektiven. Frankfurt a. M., p 131–146

Gethmann CF, Sander T (1999) Rechtfertigungsdiskurse. In: Grunwald A, Saupe S (eds) Ethik in der Technikgestaltung. Berlin, Heidelberg, New York, p 117–151

Gottschalk N, Elstner M (1997) Technik und Politik. Überlegungen zu einer innovativen Technikgestaltung. In: Elstner M (ed) Gentechnik, Ethik und Gesellschaft. Heidelberg, p 143–180

Grin J, Graaf H van de, Hoppe R (1997) Technology Assessment through Interaction. Amsterdam

Hennen L (1999) Partizipation und Technikfolgenabschätzung. In: Bröchler S, Simonis G, Sundermann K (eds) Handbuch Technikfolgenabschätzung. Berlin, p 565 –571

Horkheimer M, Adorno TW (1988) Dialektik der Aufklärung. Philosophische Fragmente. Frankfurt a. M.

Janich P, Kambartel F, Mittelstraß J (1974) Wissenschaft als Wissenschaftskritik. Frankfurt a. M.

Kant I (1907) Metaphysik der Sitten. In: Kants Werke VI. Berlin, p 203–494

Kant I (1912a) Beantwortung der Frage: Was ist Aufklärung? In: Kants Werke VIII. Berlin, p 33–42

Kant I (1912b) Zum ewigen Frieden. In: Kants Werke VIII. Berlin, p 341–386

Kass G (2000) Recent Developments in Public Participation in the United Kingdom. In: TA-Datenbank-Nachrichten Nr. 3, p 20–28

Klüver L (1995) Consensus Conferences at the Danish Board of Technology. In: Joss S, Durant J (eds) Participation in Science – the role of consensus conferences in Europe. London, p 41–49

Köberle S, Gloede F, Hennen L (eds) (1997) Diskursive Verständigung? Mediation und Partizipation in Technikkontroversen. Baden-Baden

Krüger H (1966) Allgemeine Staatslehre, Stuttgart u.a. 2. Auflage

Lauritzen P (2000) Cloning and the Future of Human Embryo Research. Oxford

Marx K, Engels F (1969) Die deutsche Ideologie. Kritik der neuesten deutschen Philosophie in ihren Repräsentanten Feuerbach, B. Bauer und Stirner und des deutschen Sozialismus in seinen verschiedenen Propheten. In: Karl Marx – Friedrich Engels – Werke, Vl. 3, Berlin (east) p 5–530

Meyer G (1999) Communication about Risk: Let Laymen lay the Foundations. Project Info Nr. 127 April 1999 from the Danish Board of Technology. Copenhagen

Rip A, Misa TJ, Schot JW (eds) (1995) Managing Technology in Society. The Approach of Constructive Technology Assessment. London

Schot W (1992) Constructive Technology Assessment and Technology Dynamics: The Case of Clean Technologies. In: Science, Technology and Human Values 17, p 36–56

Singer P, Kuhse H, Buckle S, Dawson K, Kasimba P (1993) Embryo Experimentation. Ethical, Legal and Social Issues, Cambridge

2 Post-Normal Science. Science and Governance under Conditions of Complexity

Silvio Funtowicz and Jerry Ravetz

2.1
Introduction

Policy-related research issues are particularly challenging for science. They include masses of detail concerning many particular topics, which require separate analysis and management. At the same time, there are broad strategic considerations which should guide regulatory work, such as those connected with precaution, safety or sustainability. Nothing can be managed in a convenient isolation; problems are mutually implicated extending across many scale levels of space and time; and uncertainties and value-loadings of all sorts and all degrees of severity affect data and theories alike.

This situation is a novel one for policy makers. In one sense risks and safety are in the domain of science: the phenomena of concern are located in the world of nature. Yet the tasks are totally different from those traditionally conceived for Western science. For that, it was a matter of conquest and control of nature; now we must manage, accommodate and adjust. We know that we are no longer, and never really were, the "masters and possessors of Nature" that Descartes (1824) imagined for our role in the world.

To engage in these new tasks we need new intellectual tools. A picture of reality designed for controlled experimentation and abstract theory building can be very effective with complex phenomena reduced to their simple, atomic elements. But it is not best suited for the tasks of science and governance today. The scientific mind-set fosters expectations of regularity, simplicity and certainty in the phenomena and in our interventions. But these can inhibit the growth of our understanding of the problems and of appropriate methods to their solution. Here we shall introduce and articulate several concepts which can provide elements of a framework to understand these issues. They are all new, and still evolving. There is no orthodoxy concerning their content or the conditions of their application.

The leading concept is "complexity". This relates to the structure and properties of the phenomena and the issues for policy. Systems that are complex are not merely complicated; by their nature they involve deep uncertainties and a plurality of legitimate perspectives. Hence the methodologies of traditional laboratory-based science are of restricted effectiveness in this new context.

An approach for managing complex policy-related science issues is "Post-Normal Science" (Funtowicz and Ravetz 1992, 1993; Futures 1999). This focuses on aspects of problem solving that tend to be neglected in traditional accounts of scientific practice: uncertainty and value loading. It provides a coherent explanation of the need for greater participation in science-policy processes, based on the new tasks of quality assurance in these problem-areas.

2.2
Complexity

Anyone trying to comprehend policy-related research might well be bewildered by the number, variety and complication of the issues. There is a natural temptation to try to reduce them to simpler, more manageable elements, as with mathematical models and computer simulations. This, after all, has been the successful programme of Western science and technology up to now. But the problems here have features which prevent reductionist approaches from having any, but the most limited useful effect. These are what we mean when we use the term "complexity".

Complexity is a property of certain sorts of systems; it distinguishes them from those that are simple, or merely complicated. Simple systems can be captured (in theory or in practice) by a deterministic, linear causal analysis. Such are the classic scientific explanations, notably those of high-prestige fields like mathematical physics. Sometimes such a system requires more variables for its explanation or control than can be neatly managed in its theory. Then the task is accomplished by other methods; and the system is "complicated". The distinction between science and engineering, the latter occurring when more than half a dozen variables are in play, is a good example of the distinction between simple and complicated systems.

With true complexity, we are dealing with phenomena of a different sort. There are many definitions of complexity, all overlapping, deriving from the various areas of scientific practice with, for example, ecological systems, organisms, social institutions, or the "artificial" simulations of any of them. Here we adopt a more general approach to the concept. First, we think of a "system", a collection of elements and subsystems, defined by their relations within some sort of hierarchy or hierarchies. The hierarchy may be one of inclusion and scale, as in an ecosystem with (say) a pond, its stream, the watershed, and the region, at ascending levels. Or it may be a hierarchy of function, as in an organism and its separate organs. A species and its individual members form a system with hierarchies of both inclusion and function. Environmental and technological systems may also include human and institutional sub-systems, which are themselves systems. These latter are a very special sort of system which we call "reflexive". In those, the elements have purposes of their own which they may attempt to achieve independently of, or even in opposition to, their assigned functions in the hierarchy (Funtowicz and Ravetz 1997).

First, any "system" is itself an intellectual construct that some humans have imposed on a set of phenomena and their explanations. Sometimes it is convenient to leave the observer out of the system; but in the cases of systems with human and institutional components, this is counterproductive. For environmental systems, then, the observer and analyst are there, as embedded in their own systems, variously social, geographical and cognitive. For policy purposes, a very basic property of observed and analysed complex systems might be called "feeling the elephant", after the Indian fable of the five blind men trying to guess the object they were touching by feeling a part of an elephant. Each conceived the object after his own partial imaging process (the leg indicated a tree, the side a wall, the trunk a snake, etc.); it was left to an outside observer to visualise the whole elephant. This parable reminds us that every observer and analyst of a complex system operates with certain criteria of selection of phenomena, at a certain scale-level, and with certain built-in values and commitments. The result of their separate observations and analyses are not at all "purely subjective" or arbitrary; but none of them singly can encompass the whole system. Looking at the process as a whole, we may ask whether an awareness of their own limitations is built into their personal systematic understanding, or whether it is excluded. In the absence of such awareness, we have old-fashioned technical expertise; when analysis is enriched by its presence, we have Post-Normal Science.

We can express the point in a somewhat more systematic fashion, in terms of two key properties of complex systems. One is the presence of significant and irreducible uncertainties of various sorts in any analysis; and the other is a multiplicity of legitimate perspectives on any problem. For the uncertainty we have a sort of "Heisenberg effect" where the acts of observation and analysis become part of the activity of the system under study and so influence it in various ways. This is well known in reflexive social systems, through the phenomena of "moral hazard", self-fulfilling prophecies and mass panic.

But there is another cause of uncertainty, more characteristic of complex systems. This derives from the fact that any analysis (and indeed any observation) must deal with an artificial, usually truncated system. The concepts in whose terms existing data is organised will only accidentally coincide with the boundaries and structures that are relevant to a given policy issue. Thus, social and environmental statistics are usually available (if at all) in aggregations created by governments with other problems in mind; they need interpreting or massaging to make them relevant to the problem at hand. Along with their obvious, technical uncertainties resulting from the operations of data collection and aggregation, the data will have deeper, structural uncertainties, not amenable to quantitative analysis, which may actually be decisive for the quality of the information being presented.

A similar analysis yields the conclusion that there is no unique, privileged perspective on the system. The criteria for selection of data, truncation of models, and formation of theoretical constructs are value-laden, and the values are those embodied in the societal or institutional system in which the science is being done. This is not a proclamation of "relativism" or anarchy. Rather, it is a reminder that the decision process on policies must include dialogue among those who have an interest in the issue and a commitment to its solution. It also suggests that the process towards a decision may be as important as the details of the decision that is finally achieved.

For an example of this plurality of perspectives, we may imagine a group of people gazing at a hillside. One of them "sees" a particular sort of forest, another an archaeological site; another a potential suburb, yet another sees a planning problem. Each uses their training to evaluate what they see, in relation to their tasks. Their perceptions are conditioned by a variety of structures, cognitive and institutional, with both explicit and tacit elements. In a policy process, their separate visions may well come into conflict, and some stakeholders may even deny the legitimacy of the commitments and the validity of the perceptions of others. Each perceives his or her own elephant, as it were. The task of the facilitator is to see those partial systems from a broader perspective, and to find or create some overlap among them all, so that there can be agreement or at least acquiescence in a policy. For those who have this integrating task, it helps to understand that this diversity and possible conflict is not an unfortunate accident that could be eliminated by better natural or social science. It is inherent to the character of the complex system that is realised in that particular hillside.

These two key properties of complex systems, radical uncertainty and plurality of legitimate perspectives, help to define the programme. They show why policy cannot be shaped around the idealised linear path of the gathering and then the application of scientific knowledge. Rather, the formation of policy is itself embedded as a subsystem in the total complex system of which the problem at hand is just another element.

2.3
Post-Normal Science as a bridge between complex systems and policy

The idea of a science being somehow "post-normal" conveys an air of paradox and perhaps mystery. By "normality" we mean two things. One is the picture of research science as "normally" consisting of puzzle solving within an unquestioned and unquestionable "paradigm", in the theory of Thomas Kuhn (1962). Another is the assumption that the policy context is still "normal", in that such routine puzzle solving by experts provides an adequate knowledge base for policy decisions. Of course researchers and experts must do routine work on small-scale problems; the question is how the framework is set, by whom, and with whose awareness of the process. In "normality", either science or policy, the process is managed largely implicitly, and is accepted unwittingly by all who wish to join in. The great lesson of recent years is that that assumption no longer holds. We may call it a "post-modern rejection of grand narratives", or a green, NIMBY (Not In My Back Yard) politics. Whatever its causes, we can no longer assume the presence of this sort of "normality" of the policy process.

The insight leading to Post-Normal Science is that in the sorts of issue-driven problems characteristic of policy-related research, typically facts are uncertain, values in dispute, stakes high, and decisions urgent. Some might say that such problems should not be called "science"; but the answer could be that such problems are everywhere, and when science is (as it must be) applied to them, the conditions are anything but "normal". For the previous distinction between "hard", objective scientific facts and "soft", subjective value-judgements is now inverted. All too often, we must make hard policy decisions where our only scientific inputs are irremediably soft.

The difference between old and new conditions can be shown by the present difficulties of the classical economics' approach to environmental policy. Traditionally, economics attempted to show how social goals could be best achieved by means of mechanisms operating automatically, in an essentially simple system. The "hidden hand" metaphor of Adam Smith conveyed the idea that conscious interference in the workings of the economic system would do no good and much harm; and this view has persisted from then to now. But for the achievement of sustainability, automatic mechanisms are clearly insufficient. Even when pricing rather than control is used for implementation of economic policies, the prices must be set, consciously, by some agency; and this is then a highly visible controlling hand. When externalities are uncertain and irreversible, then no one can set "ecologically correct prices" practised in actual markets or in fictitious markets (through contingent valuation or other economic techniques). There might at best be "ecologically corrected prices", set by a decision-making system. The hypotheses, theories, visions and prejudices of the policy-setting agents are then in play, sometimes quite publicly so. And the public also sees contrasting and conflicting visions among those in the policy arena, all of which are plausible and none of which admits of refutation by any other. This is a social system, which, in the terms discussed above, is truly complex, indeed reflexively complex.

In such contexts of complexity, there is a new role for science. The facts that are taught from textbooks in institutions are still necessary, but are no longer sufficient. For these relate to a standardised version of the world, frequently to the artificially pure and stable conditions of a laboratory experiment. The world as we interact with it is quite different. Those who have become accredited experts through a course of academic study have much valuable knowledge in relation to these practical problems. But they may also need to recover from the mindset they might absorb unconsciously from their instruction. Contrary to the impression conveyed by textbooks, most problems in practice have more than one plausible answer; and many have no answer at all.

Further, in the artificial world studied in academic courses, it is strictly inconceivable that problems could be tackled and solved except by deploying the accredited expertise. Systems of management of problems that do not involve science, and which cannot be immediately explained on scientific principles, are commonly dismissed as the products of blind tradition or chance. And when persons with no formal qualifications attempt to participate in the processes of innovation, evaluation or decision, their efforts are viewed with scorn or suspicion. Such attitudes do not arise from malevolence; they are inevitable products of a scientific training which presupposes and then indoctrinates the assumption that all problems are simple and scientific, to be solved on the analogy of the textbook.

It is when the textbook analogy fails, that science in the policy context must become post-normal. When the assumptions of simplicity and certainty are totally inappropriate, the goal of achievement of factual knowledge must be substantially modified. In post-normal conditions, aiming at hard, objective facts may be an irrelevance, indeed a diversion. Here, the guiding principle is a more robust one, that of quality.

It could well be argued that quality has always been the effective principle in practical research science, but it was largely ignored by the dominant philosophy and ideology of science. For post-normal science, quality becomes crucial, and quality refers to process at least as much as to product. It is increasingly realised in policy circles that in complex issues, lacking neat solutions and requiring support from all stakeholders, the quality of the decision-making process is absolutely critical for the achievement of an effective product in the decision. This new understanding applies to the scientific aspect of decision-making as much as to any other.

Post-Normal Science can be located in relation to the more traditional complementary strategies, by means of a diagram (see Figure 2.3). On it, we see two axes, "systems uncertainties" and "decision stakes". When both are small, we are in the realm of "normal", safe science, where expertise is fully effective. When either is medium, then the application of routine techniques is not enough; skill, judgement, sometimes even courage are required. We call this "professional consultancy", with the examples of the surgeon or the senior engineer in mind. Our modern society has depended on armies of "applied scientists" pushing forward the frontiers of knowledge and technique, with the professionals performing an aristocratic role, either as innovators or as guardians.

Of course there have always been problems that science could not solve; indeed, the great achievement of our civilisation has been to tame nature in so many ways, so that for unprecedented numbers of people, life is more safe, convenient and comfortable than could ever have been imagined in earlier times. But now we are finding that the conquest of nature is not complete. As we confront nature in its reactive state, we find extreme uncertainties in our understanding of its complex systems, uncertainties that will not be resolved by mere growth in our databases or computing power. And since we are all involved with managing the natural world to our personal and sectional advantage, any policy for change is bound to affect our interests. Hence in any problem-solving strategy, the decision-stakes of the various stakeholders must also be reckoned with.

This is why the diagram has two dimensions; this is an innovation for descriptions of science, which had traditionally been assumed to be value-free. But in any real problem of policy-related research, the two dimensions are inseparable. When conclusions are not completely determined by the scientific facts, inferences will (naturally and legitimately) be conditioned by the values held by the agent. This is a necessary part of ordinary research practice; all statistical tests have values built in through the choice of numerical "confidence limits", and the management of "outlier" data calls for judgements that can sometimes approach the post-normal in their complexity. If the stakes are very high (as when an institution is seriously threatened by a policy) then a defensive policy will involve challenging every step of a scientific argument, even if the systems uncertainties are actually small. Such tactics become wrong only when they are conducted covertly, as by scientists who present themselves as impartial judges when they are actually committed advocates. There are now many initiatives, increasing in number and significance all the time, for involving wider circles of people in decision-making and implementation on environmental issues.

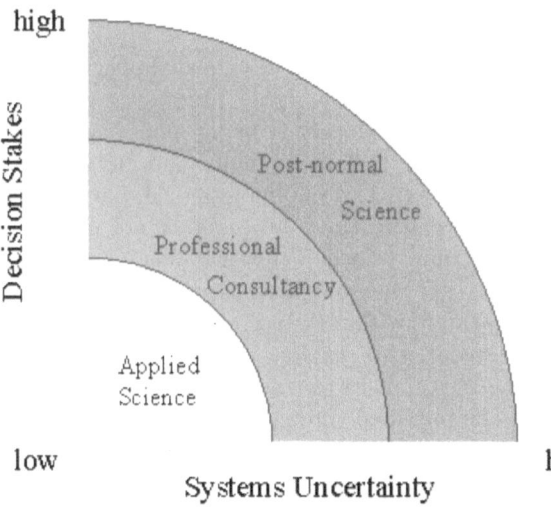

The contribution of all the stakeholders in cases of Post-Normal Science is not merely a matter of broader democratic participation. For these new problems are in many ways different from those of research science, professional practice, or industrial development. Each of those has its means for quality assurance of the products of the work, be they peer review, professional associations, or the market. For these new problems, quality depends on open dialogue between all those affected. This we call an "extended peer community", consisting not merely of persons with some form or other of institutional accreditation, but rather of all those with a desire to participate in the resolution of the issue. Seen out of context, such a proposal might seem to involve a dilution of the authority of science, and its dragging into the arena of politics. But we are here not talking about the traditional areas of research and industrial development; but about those where issues of quality are crucial, and traditional mechanisms of quality assurance are patently inadequate. Since this context of science is one involving policy, we might see this extension of peer communities as analogous to earlier extensions of franchise in other fields, as allowing workers to form trade unions and women to vote. In all such cases, there were prophecies of doom, which were not realised.

For the formation of policy under conditions of complexity, it is hard to imagine any viable alternative to extended peer communities. They are already being created, in increasing numbers, either when the authorities cannot see a way forward, or know that without a broad base of consensus, no policies can succeed. They are called "citizens' juries", "focus groups", or "consensus conferences", or any one of a great variety of names; and their forms and powers are correspondingly varied. But they all have one important element in common: they assess the quality of policy proposals, including a scientific element, on the basis of whatever science they can master during the preparation period. And their verdicts all have some degree of moral force and hence political influence.

Along with this regulatory, evaluative function of extended peer communities, another, more intimately involved in the policy process, is springing up. Particularly at the local level, the discovery is being made, again and again, that people not only care about their community and environment but also can become ingenious and creative in finding practical, partly technological, ways towards their improvement. Here the quality is not merely in the verification, but also in the creation; as local people can imagine solutions and reformulate problems in ways that the accredited experts, with the best will in the world, do not find "normal" within their professional paradigms.

None can claim that the restoration of quality through extended peer communities will occur easily, and without its own sorts of errors. But in the processes of extension of peer communities through the approach of Post-Normal Science, we can see a way forward, for science as much as for the complex problems of policy-related research.

A sort of manual for Post-Normal Science practice has been produced by the UK Royal Commission on Environmental Pollution (1998). In its 21st Report, on Setting Environmental Standards, it makes a number of observations and recommendations reflecting this new understanding. Thus, on uncertainty, we have:

- 9.49: No satisfactory way has been devised of measuring risk to the natural environment, even in principle, let alone defining what scale of risk should be regarded as tolerable;

on values:

- 9.74: When environmental standards are set or other judgements made about environmental issues, decisions must be informed by an understanding of peoples' values. ... ;

and on extended peer communities:

- 9.74 (continued): Traditional forms of consultation, while they have provided useful insights, are not an adequate method of articulating values;

and on a plurality of legitimate perspectives:

- 9.76: A more rigorous and wide-ranging exploration of people's values requires discussion and debate to allow a range of viewpoints and perspectives to be considered, and individual values developed.

2.4
Conclusion

The inadequacies of the traditional "normal science" approach have been revealed with dramatic clarity in the episode of "mad cow" disease. For years the accredited researchers and advisors assured the British government that the likelihood of transfer of the infective agent to humans was very small. They did not remark on the sorts of decision-stakes involved in the policy, in which public alarm and government expense were the main perceived dangers. The risk of an epidemic among humans (with its attendant costs) was discounted by the experts and eventually officially denied. When the human cases of neo-variant CJD were confirmed and related to CJD, it was admitted by both experts and officials that an epidemic of this degenerative disease was a "non-quantifiable risk". The situation went out of control, and the revulsion of consumers threatened not only British beef, but also perhaps the entire European meat industry. At this stage there had to be a "hard" decision to be taken, on the number of cattle to be destroyed, whose basis was a very "soft" estimate of how many cattle deaths would be needed to reassure the meat-eating public. At the same time, independent critics who had been dealt with quite harshly in the past were admitted into the dialogue. Without in any way desiring such an outcome, the British Ministry of Agriculture, Forests and Fisheries had created a situation of extreme systems uncertainty, vast decision stakes, and a legitimated extended peer community.

The Post-Normal Science approach needs not be interpreted as an attack on the accredited experts, but rather as assistance. The world of "normal science" in which they were trained has its place in policy-related research, but it needs to be supplemented by awareness of the "post-normal" nature of the problems we now confront. The management of complex natural systems as if they were simple scientific exercises has brought us to our present mixture of triumph and peril. We are now witnessing the emergence of a new approach to problem-solving strategies in which the role of science, still essential, is now appreciated in its full context of the uncertainties of natural systems and the relevance of human values.

References

Descartes R (1824) Discours de la méthode, Part VI. Édition Victor Cousin Paris

Futures (1999) Special Issue: Post-Normal Science. Ravetz JR (ed) 31:7.

Funtowicz SO, Ravetz JR (1992) Three Types of Risk Assessment and the Emergence of Post-Normal Science. In: Krimsky S, Golding D (eds) Social Theories of Risk. Praeger, Westport (CN), pp 251–273

Funtowicz SO, Ravetz JR (1993) Science for the post-normal age. Futures 25:7: 739–755

Funtowicz SO, Ravetz JR (1997) The Poetry of Thermodynamics. Futures 29:9: 791–810

Kuhn TS (1962) The Structure of the Scientific Revolutions. University of Chicago Press, Chicago, IL

UK Royal Commission on Environmental Pollution (1998) Setting Environmental Standards. 21st Report, Chapter 9 – Conclusions
[http://www.rcep.org.uk/]

3   Business as Usual: On the Prospects of Normality in Scientific Research

Martin Carrier

In a series of influential papers composed together with Jerry Ravetz, Silvio Funtowicz attempts to sketch the methodology of a new type of science: Post-Normal Science. To state it right at the start: I'm not convinced. I strongly doubt that Funtowicz has managed to provide categories that are suitable for singling out a significant, new phenomenon. That is, I wish to make three claims:

1. The collection of properties he lists fails to capture the essential characteristics of policy-related research.
2. These properties do not affect the structure of science; no methodological issues are concerned here.
3. The suggested restrictions are by no means historically novel; rather, they have been part of science since its inception.[1]

In sum, there is no need to venture a methodological turn toward "Post-Normal" Science. In this area, the prospects of normality aren't that bad. Business as usual can help – at least to some extent.

3.1
Lack of selectivity of the suggested features

Policy-related research issues are claimed to exhibit a particular, "post-normal" structure. This conception is deliberately fashioned as a contrast to Kuhn's normal science. Post-Normal Science is said to give up the commitment to a single paradigm for guiding research, and it also breaks with the idea that expert knowledge gained within such a restricted approach is suitable for policy advice. Rather, Post-Normal Science is assumed to grapple with uncertain facts, disputed values, high epistemic, ethical or economic stakes, and the urgent need for decisions (see chap. 2, p 19).

Funtowicz provides a second characterization of post-normality that is inconsistent with the one given. According to the first notion, post-normality

---

[1] Somewhat surprisingly, Funtowicz begins by saying that the "situation is a novel one for policy makers" (see chap. 2, p 15) but then goes on to explain the differences between traditional *science* and present-day policy-related *research*.

demands that all the four features mentioned are present. This conjunctive reading is confirmed by Funtowicz and Ravetz (1994, 1882) where they are listed using the connective "and". In contrast, Fig. 2.3 entails that high scores on either one of the relevant dimensions suffice for the emergence of Post-Normal Science (see chap. 2, p 21). This disjunctive reading is buttressed by Funtowicz and Ravetz (1994, 1881), where Post-Normal Science is linked to the prevalence of high decision stakes "or" high system uncertainties. That is, the first concept of post-normality is much stronger than the second. The strong concept is based on the conjunction of four criteria whereas the weak one involves the disjunction of only two of them. The weak concept doesn't appear overly significant. It would imply that simply spending a lot of money on a project would make this endeavor post-normal – which appears wildly implausible. Even a conjunctive reading of the two criteria that are consti-tutive of the weak concept doesn't help. Hazardous betting, or waging war for that matter, are characterized by both high uncertainty and high stakes. But neither activity strikes me as particularly post-normal. For this reason I disregard the second route to post-normality and focus on the strong concept, i. e., the conjunction of uncertainty, value disputes, high stakes and urgency.

It helps to bring things down to earth if one realizes that only a tiny frac-tion of science is at all affected by these features. First, exclusively application-oriented science is considered relevant; science as an epistemic enterprise is explicitly left out of the picture. This neglect concerns innovative "revolu-tionary" science as well as routine research which is conceived in a quite traditional Kuhnian sense (1994, 1882–1883).[2] Second, within the realm of application-oriented science, the vast majority of scientific endeavors is not covered either. Look at the following technological challenges: development of energy-saving techniques (such as the three-litres automobile), further increase in the integration and efficiency of electronic devices (including projects such as molecular circuits and switches or issues like quantum com-putation), construction of the Airbus 3XX. Technological innovations of this sort will shape our future. But neither is any outstanding degree of uncer-tainty involved here, nor are values disputed nor urgent decisions called for. So, the dominant and most widespread type of applied research remains un-affected by the methodological reorientation that Funtowicz envisages.

Actually, this neglect of wide areas of present-day applied research seems to me a major liability of Funtowicz's approach, since it is at this juncture where important methodological challenges emerge. Much of this research

---

[2] Funtowicz's strong inclination toward Kuhn as his model in philosophy of science gives his conception a somewhat old-fashioned ring. Kuhn's model has been chal-lenged since its enunciation and is now largely replaced by different views. Most of these later approaches stress theoretical pluralism which is one of the distinc-tions of Post-Normal Science as well. In this respect, Funtowicz's allegedly new claims are actually "middle-of-the-road views" within present-day philosophy of science.

proceeds in an area located between pure and applied science. Such projects are largely intended to produce marketable goods, for one, but they transcend the mere application of pre-existing, comprehensive theories, for another. Rather, in such application-dominated research, as I call it, theoretical innovations are developed with practical purposes in mind. And it is worthwhile to examine what the preponderance of application-dominated research does to scientific methodology. However, Funtowicz quite conservatively reproduces the traditional distinction between pure and applied science.

Third, even large areas of policy-related research (which is at his primary focus) are not captured by the criteria Funtowicz advances. Consider the depletion of the ozone layer which is most certainly relevant to policy advice. But there is no particular uncertainty involved. The causes, the mechanism, and the effects are completely transparent, and the causal chain was clarified by doing business as usual, namely, traditional chemical research. Judging by Funtowicz's lights, nothing post-normal should be present here.

Global warming, by contrast, exhibits the kind of uncertainty Funtowicz deems central to post-normality. Nobody is able to predict precisely the future impact of greenhouse gases. And it goes without saying that stakes are high and decisions urgent in this area. So we are dealing with a clear-cut example of Post-Normal Science. But it seems to me that there is no significant methodological distinction between the ozone and the greenhouse problems. The latter may be more difficult to solve but I don't see why hard problems should demand a different methodology. Funtowicz's criteria introduce divisions that appear implausible to me; they separate what should be kept together.

Consider, in opposition, the issue of overhauling the European social security systems. The relevant attempts are subject to a large number of uncertainties. We are ignorant of the future rise of medical cost, we neither know the development of the age structure of the population (because of uncertain immigration figures) nor the magnitude of economic growth (which is influential on the level of affordable medical expenses). In addition to uncertainties concerning the factual basis, the decisions made are clearly determined by moral values, and they are urgent. Judging by Funtowicz's criteria, social security reform comes out as a piece of Post-Normal Science. But this seems absurd to me. Science is only superficially concerned with regulating this issue. It's an administrative or institutional problem, not a scientific one. Funtowicz's criteria lump together what should be kept separate. On the whole, then, his criteria are not sufficiently selective.

3.2
Methodological irrelevance of the listed properties

It is true that science is heavily involved in the business of advising politicians. Still, it seems urgent to distinguish more clearly between characteristics having to do with science proper and political features of the decision-making process. To be sure, some political processes are informed by scientific information; but this does not make them scientific processes. The improvement of knowledge by observation, experiment, and theory formation is one thing; the mediation of conflicting interests or the raise of support for one's projects is quite another thing. The mishandling of the BSE problem Funtowicz describes (see chap. 2, p 23) is a glaring management failure, caused by an inappropriate weighting of interests. As a result, the political maneuvers backfired and measures had to be taken in order to placate the public. It is not scientific uncertainty in itself, but the politicians' vain attempt to stage a cover-up operation that brought about the political disaster. Politics, not science, was primarily responsible for letting the situation get out of hand.

The same lack of distinction between science and politics underlies another one of Funtowicz's theses, namely, the claim of a particular value-ladenness of Post-Normal Science. Look at the argument (see chap. 2, p 21): It is said, first, that any policy dealing with the natural world in order to promote some human interests has to take into account the relevant "decision stakes", i. e., economic interests or ethical concerns. This is obviously true. But Funtowicz suggests the conclusion that science ceases to be value-free – which in no way follows. The postulate that science should be value-free exclusively rules out that science enunciates or assesses ethical obligations in its own right. Epistemic values are not concerned, nor is science barred from addressing value judgments passed by others. Funtowicz mentions the conventional threshold values set in the application of statistical tests. Such considerations concern the issue as to how well a hypothesis is taken to be confirmed by the available evidence. Epistemic evaluations of this sort have been part of standard philosophy of science for decades. There is nothing specifically post-normal in this realm.

Second, it is true that some instances of research are intimately intertwined with non-cognitive values. Examinations of the security of a motor car or of the reliability of an electric lawn-mower are cases in point. It is part of the rationale of such investigations that security and reliability are worthwhile. But value-ladenness of this sort is a rather trivial feature for two reasons. It does not involve an influence of ethical values on the results of the investigation. These commitments only provide the motives for performing the study. Moreover, the pursuit of such an examination does not commit the scientist herself to any particular values. The security of a car might as well be assessed by the evil scientist, eager to promote the decline of humankind. The evil one aims to boost the sales of the most insecure cars and the most unreliable lawn-mowers, and one of the conditions of success in this repugnant endeavor is the trustworthy appreciation of security and reliability levels. Ethical commitments are no precondition for doing such research and they are unconnected to its outcome. I conclude that no persuasive argument has been ventured that could justify the inherent value-ladenness of application-oriented science in general and of Post-Normal Science in particular.

3.3
The perpetuity of Post-Normal Science

It is doubtlessly true that the significance of science for policy advice has increased drastically in the course of the 20th century. Consider, by way of contrast, the anecdote attributed to President Woodrow Wilson. When the U.S. entered the First World War, Wilson pondered the composition of a body of scientists requested to advise the political leaders. He decided to appoint one physicist to serve in this body "in case we have to calculate something." Wilson's justification for this appointment highlights the change in the assessment of the political relevance of the sciences. In this respect, I agree with Funtowicz: the import of policy-related research has increased dramatically during the past century.

However, this is a matter of degree, not of principle. Science was brought to bear on political issues more intensively recently but the connection as such pervaded science since its inception. Consider astrology and Renaissance impetus physics. The purpose of giving astrology-based advice in practical matters, including political ones, underlay the development of astronomy in the ancient world and was also among the chief motivations for the spread of heliocentric astronomy in the 16th century. Astrology exhibits the conjunction of high uncertainty and high stakes that is advanced as being characteristic of Post-Normal Science. After all, there is a considerable chance of being mistaken in one's predictions, and the issues at stake were quite important and urgent. Decisions about war and peace were made on this basis. As a result of this insecurity we experience interventions from a wider audience, namely, the critics and the reformers of astrology. That is, we encounter the extended peer-group that is said to go along with Post-Normal Science. Finally, values are involved as well. One of the relevant issues concerned the relation between astrology and Christian faith. The conclusion is unavoidable: astrology is the first Post-Normal Science.

Likewise, one of the chief motives for attacking mechanical problems was the challenge to predict the path of canon balls. Impetus physics was pushed ahead in the 16th century so as to make such complex, uncertain situations controllable. And it goes without saying that such problem-solutions were of utmost importance for giving strategic advice to generals and warlords as to how best to pulverize enemy cities. So, unquestionably, stakes were high in this venture. One quickly realizes, in addition, that decisions about warfare are always urgent and that values are implicated as well (after all, some tension might be perceived between advancing methods for the economical slaughtering of citizens and the commandment to love one's neighbor). Impetus mechanics comes out as a piece of Post-Normal Science. The upshot is that science as a whole, and modern science in particular, got underway as policy-related, post-normal research.

Second, and more importantly, Funtowicz conveys the impression that present-day science fares particularly badly in dealing with policy-related research issues. Traditional Western science "was a matter of conquest and control of nature; now we must manage, accommodate and adjust" (see chap. 2, p 15).[3] This suggests that it is only in our time that the restrictions of science have come to light. We find *now* that the conquest of nature is not complete (see chap. 2, p 21). Science is said to enter the post-normal era which ostensibly involves uncertainty bordering at ignorance and uncontrollability. But in fact, the reverse is true. The success of science in mastering complex phenomena has increased drastically. Restrictions in predictability were much more marked in previous ages. Astrology-supported advice and impetus-based military operations should have been less than successful. Analogously, Christopher Wren was well-acquainted with the then recent Newtonian mechanics. But when he set out to construct St. Paul's Cathedral he relied completely on traditional medieval rules. And quite understandably so. For although classical mechanics was tremendously successful in elucidating the fundamental issues of mechanics, it was of little help in practical problems like the one Wren was facing.

Through the ages science could never deliver on the promises of comprehensive explanation and control. But the capacity of science to capture complex phenomena has expanded during history. Since the dawn of curiosity, science operates at the verge of overtaxing itself. This is what keeps progress underway. It is a fact that today science manages to meet complex challenges it couldn't even remotely address earlier. To be sure, knowledge is still limited, and always will remain limited at that; but this should not mislead us into thinking that science is somehow losing grip on complexity. Quite the contrary. The false impression of insufficiency is created by the constant enlargement of the purview of science and the incessant rise of expectations tied to it. Science can now achieve more than it ever could. But the demands brought to science have risen even faster. This is what created the misleading impression that science has to go politics when being faced with complex phenomena.

3.4
The non-existence of normal science

In the preceding section I argued that science has been more post-normal in the course of its development than Funtowicz believes. Now I wish to make the converse claim, namely, that past science has been much less "normal".

---

[3] Astonishingly enough, Funtowicz may be interpreted to the effect that he revokes this very claim of distinctiveness immediately after the quoted passage. There he admits that we have never really been the masters of nature (ibid.).

The standard situation from which Post-Normal Science is said to deviate is described by two characteristics. First, as to science proper, we encounter a Kuhnian-structured normal science, i. e., the application of a received view, a paradigm, whose principles are not called into question and are not modified or developed in any significant respect. Second, as to politics, knowledge gained in this way provides an adequate factual basis for policy decisions (see chap. 2, p 19). I doubt that a situation of this sort has ever existed.

The one discipline that is most extensively used in policy advice is economics. And yet it doesn't fit the bill. Economics during the past half century was not governed by an intellectual monopoly but exhibited an ongoing rivalry between so-called neo-classic and Keynsean approaches. And it would come as a surprise to me if textbook knowledge in economics had ever been sufficient for guiding policy decisions. So, even in this standard situation none of Funtowicz's conditions is actually met. This suggests that even traditional scientific approaches go beyond of what Funtowicz deems normal. I suspect that normal science in his twofold sense has never existed at all.

References

Funtowicz S O, Ravetz J R (1994) Uncertainty, Complexity and Post-Normal Science. In: Experimental Toxicology and Chemistry 13: 1881–1885.

4 Rational Technology Assessment as Interdisciplinary Research

Michael Decker and Armin Grunwald

4.1
Overview

The concept of rational Technology Assessment has been created and developed by an interdisciplinary scientific team of the Europäische Akademie.[1] Several aspects of this approach to Technology Assessment have been published in the meantime (Decker 2000, Grunwald 1999, Grunwald/Saupe 1999, Grunwald 2000a, Grunwald 2000b). While Rational Technology Assessment is, up to now, mainly focussed on conceptual and methodological questions (in order to clarify the differences compared to other approaches to TA, cf. Grunwald 1999, pp 11–28), the intention of the present paper is to relate the organisation principle of the interdisciplinary research done by the Europäische Akademie to the methodological and conceptual cernel of rational TA. Accordingly, the paper starts by explaining central issues of rational TA, such as the concept of problem-oriented research and the concept of pragmatic rationality, and by deriving the resulting requirements for the quality control of rational TA as interdisciplinary research. The second half of the paper is dedicated to the way by which these requirements are realised in accordance with the organisation principles of the Europäische Akademie, namely to bring together interdisciplinary project groups consisting of disciplinary experts. In this way, the paper aims at interrelating theory and practice of Technology Assessment as interdisciplinary research as well as to discuss the chances of the approach presented to overcome the well-known expert dilemmas.

4.2
Technology Assessment as problem-oriented research

Technology Assessment is currently often considered as an instrument to allow or to enhance social mediation processes concerning decision-making

---

[1] The main contributors have been Michael Decker, Carl Friedrich Gethmann, Armin Grunwald, Mathias Gutmann and Gerd Hanekamp.

on technology options (e. g. Zweck 1993). The role of scientific research in these processes of mediation, compromising or problem-solving often seems to be only marginal. Such trends, however, are not followed in Rational TA. Though science indeed has no privileged access in defining the "common good" and has only the legitimisation to give advice to decision-makers and to the public instead to influence or to make extra-scientific decisions, there are many reasons for not understating the role of scientific research in the processes of social shaping of technology. The main reason, of course, is the lack of knowledge in many fields concerning the impacts and consequences of technologies: actual developments of the use of the internet, regulation requirements in order to prevent misuse of the internet, operationalisation of the concept of sustainability for social activity fields, the potentials of key technologies in contributing to sustainable development are only some examples.

One major objection against TA as a primarily scientific endeavour is founded upon the so-called expert dilemmas: expertise and counter-expertise are devaluating one another, thus the trust in scientific expertise decreases, and science is suspected of being merely a stakeholder among others. In this paper it will be shown to what extent it is possible to deal with this problem by using inner-scientific means of reasoning. The main idea is to implement interactive communication between experts from different fields which are commonly dealing with a societally defined problem: problem-oriented research done by experts working together in interdisciplinary groups. In this context it seems to be appropriate to remember some key aspects of problem-oriented research.

TA is oriented to the scope of social problems and challenges related to technology: impacts and consequences of technology, political and societal ways of dealing with them, potentials for contributions to societal problem-solving and innovation policy and implementation conditions of technology are the classical fields of TA. Accordingly, TA is not governed by discipline-immanent research programmes but driven by extra-scientific problems. The reasons for a problem-oriented approach to tackle such problems lie by no means (ontologically) in the *complexity* of the subjects of the investigation or of the questions under study. Complex subjects of research can also be treated within individual disciplines, whereby each discipline studies its subjects from the standpoint of its own specific cognitive interests and by using its own special methodical approach. The fact that such methods are often inadequate for TA-related problems is due to the fact that their purpose is *to provide knowledge as a basis for acting and decision-making concerning technology and its implementation in society*. The need for integrative and interdisciplinary research on the impact and consequences of technology results from the necessity for *integrated and coherent political and societal judgement*. The scientific analysis carried out by single disciplines would be completely inadequate for a comprehensive diagnosis of the present situation, for establishing scenarios of future developments, for the elaboration and assessment of options for political action, or even for the "construction" of comprehensive political problem-solving strategies in the context of technology. Comprehensive social awareness of problems and complex situations for decision-making influencing technology presuppose making knowledge available for deciding and acting which has to extend beyond economic sectors (such as transportation, the media, the construction industry), beyond classical disciplines (such as between natural and cultural sciences), and beyond cultural preconceptions (in particular, in global challenges). The integration of knowledge from various sources takes place under problem-solving pressure. TA provides contextually-influenced and pragmatic knowledge with regard to decision-making and acting (Grunwald 2000a). These requirements lead to a lot of methodological and organisational challenges to TA as *interdisciplinary* research.

In the recent years, the concept of *transdisciplinarity* is being discussed increasingly (Mittelstraß 1998, Jaeger/Scheringer 1998). The reasons given for the necessity of transdisciplinary research seem to be the same as those which were named earlier as the reasons for interdisciplinarity: in a few words and simplified, they consist in the uncontested fact that the requirements of actual social problems cannot be dealt with within the limits of individual disciplines (Bechmann/Frederichs 1996, Mittelstraß 1998): " 'Transdisciplinarity' should be understood to mean . . . research . . . which defines and solves its problems independently of scientific disciplines" (Mittelstraß 1998, p 44, translation A. G.). Transdisciplinarity therefore means nothing else than *problem-oriented research* (Bechmann/Frederichs 1996), and refers to the non-scientific problem-orientation of its work – in other words, to the fact that research which crosses disciplinary boundaries is a means to an end, but no end in itself. At the same time, this interpretation allows us to establish criteria and standards for success or failure for this type of research: namely, to provide implementable solutions for a socially defined problem. As opposed to the concept of "interdisciplinarity", that of transdisciplinarity has the advantage that it already includes its own reasons for pursuing cross-disciplinary research – namely, that it should contribute to mastering non-scientific problems by means of scientific methods. In this paper, the notion of "interdisciplinarity" is maintained because the orientation of TA to extra-scientific problems seems to be evident.

Problem-oriented research differs in fundamental respects from classical institutionalised research (Funtowicz/Ravetz 1993,[2] Bechmann/Frederichs 1996). Its cognitive interests are not only determined by establishing knowledge of nature or of the world, as is predominantly the case in the clas-

---

[2] Cp. the paper by Silvio Funtowicz and Jerry Ravetz in this volume.

sical disciplines,[3] but rather by the problem- and decision-making orientation of the knowledge gained. The production of solid textbooks – the succession of which made the discipline's progress tangible – is no longer the goal, but rather the elaboration of project-specific results as a contribution to knowledge as a basis for decision-making and acting. Science leaves the protected niche of presumed value-freedom, takes on a politically relevant role in the definition of social problems (cp. the field of climate change), and becomes dependent – for the conditions for its success and the criteria for its quality – on its non-scientific contexts.

4.3
The Concept of Rationality in Technology Assessment

In public debate carried out in the political and academic communities on attitudes towards technology and technology impact the concept of "rationality", in contrast to such concepts as "risk", "innovation" or "acceptance", is not a central focus of attention. Due to the fact that it is no longer tenable in the modern age to understand the concept of rationality as it was understood in the period of the European Enlightenment, i.e. to consider rationality as the transcendental ability of human subjects, the conclusion was drawn that rationality could be understood only as the product of social systems, whereas one could no longer meaningfully speak of the rationality of actions and decisions (Luhmann 1990, p 692 ff.). This loss was related to the loss of authority of science in the public debate on technology, resulting from public perception of conflicts within the academic community (consider, for instance, the ongoing debate on the various kinds of expert dilemmas, e.g. Nennen/Garbe 1996).

Ambivalences in connection with the use of the concept of rationality are also reflected in political or public debate on technology impact. In actual fact, the concept of rationality is often used rhetorically or strategically to gain acceptance for one's own position. This, in turn, has opened the concept of rationality up to suspicions of ideology. It is said to be a narrowly specific and particularistic concept but at the same time claiming "imperialistically" general validity. In public debate on technology impact special reference to rationality seems to be – in certain circles – part of a specific pattern of argumentation used by *proponents* of technology.

These reservations with regard to the concept of rationality are based on two core theses which are philosophical in nature. The first of them primarily concerns mistrust, fed by the pluralistic sense of identity in present-day ("post-modern") societies, towards comprehensive integrative approaches that extend beyond pluralism. Secondly, there are considerable doubts as to

---

[3] This applies, at least, to the self-image of the classical disciplines. Methodical reconstruction, however, shows that scientific knowledge always is related to human actions within the means/ends-scheme (Janich 1992).

whether the concept of rationality has a pragmatic role to play in public debate on technology impact in the sense that its use will be able to offer practical advantages and to provide real and helpful problem-solving capacities (Grunwald 1999, pp 29–31).

In its main approaches, professional Technology Assessment (TA) has virtually ignored the question as to what criteria characterise a "rational" approach to the genesis and impacts of technology. To the extent the concept of rationality is mentioned at all, this is done mostly in the form of claiming that one of the tasks of reflection on technology impacts is that of improving rationality, for example by saying that "TA analyses are intended to raise the level of reflection and rationality of decision-makers . . ." (Paschen 1986, p. 32; translation A.G.). In statements of this kind it is always presupposed that it has been clarified consensually what is to be understood by rationality and, secondly, that rationality is to be worked towards in the process of a society formulating policies on the social shaping of technology.

The question arises why in rational TA the concept of rationality is emphasised in spite of the difficulties mentioned above. This will be explained in the following parts focussing on the relation of the notion of rationality and the idea of trans-subjective validity of propositions and requests. The central problem in TA is how to arrive at valid (descriptive) knowledge and shared (normative) orientations. In order to exhaust the possibilities of problem-oriented interdisciplinary research for the generalisation of knowledge, orientations and legitimisation the concept of rationality seems to be an appropriate umbrella to cover these different aspects.

4.3.1
The pragmatic concept of rationality

In order to justify the introduction of a concept the ends must be indicated to the overcoming of which the concept is intended to contribute in some way. Let the answer to this question be referred to as the pragmatic site of the concept of rationality. A reconstruction of the use of the concept of rationality shows that it represents a category for the assessment of actions and decisions; actions or decisions are assessed for rationality *ex ante* or *ex post*. The concept of rationality is thus a term relating to *reflection*. The purpose of the reflection on rationality is to determine whether "everyone else" in this situation would act (in the *ex ante* case) or would have acted in exactly the same way as was in fact done (in the *ex post* case). With regard to the situation of reflection *ex ante* this means assessing what the result of an action or decision would have to be, so that "everyone else" in the given situation would act in the given way or would be able to agree to the decision. This reflection leads to criteria for selecting options for acting: "Rational procedures allow us to arrive at non-arbitrary answers" (Brown 1988, p. 35). As such, and with all pragmatically necessary consideration for context dependency, this concept is to be used to reflect the generalisability of propositions and orientations with respect to individual persons. Actions and decisions are to be designated as rational when they take place on the basis of knowledge and orientations that can be made to appear reasonable or *justifiable to everyone else* (Rescher 1988, Gethmann 1995). The rationality of actions and decisions is based on the *transsubjective validatability* of statements and the justifiability of requests. The relevance of this concept to TA becomes clear in thinking about the required range of validity of knowledge and orientations determining far-ranging decisions in technology policy.

The assessment of rationality in the form reconstructed in this way takes place from the *perspective of participants* in ongoing communication. The basis of a deliberative assessment of this kind can thus only be a *predeliberative agreement* that is shared in this community (Grunwald 2000a).[4] Rationality is always constituted on the basis of a community of players. Neither references to systems rationality (Luhmann 1990) nor an exclusive focus on the examination of isolated players as is the case in some economic rationality models can be justified on the basis of this pragmatic view. Rationality is always "ours"; whether anyone will consider our decisions to be system-rational a hundred years from now may be an interesting historical question when the time comes; but it is totally irrelevant for our present actions and decision-making.

Rationality including the definition of the criteria for rationality appears to be a *self-construct of society* in order to operationalise the regulative idea of trans-subjective validity of knowledge and orientations. It must be created in the face of the immanence of society and cannot be derived from first principles. This means that rationality criteria based on understandability "for everyone" must have their basis in the factual elements of society, at least if Technology Assessments that are constituted upon them shall to be commonly accepted and practically implemented. Because they have a "site in life" criteria of this kind need a basis in the factual elements of society, itself culturally variant – but far from being arbitrary (Grunwald 1999). This interrelationship between acceptance and acceptability, between the factual and the counterfactual, between continuity and change constitutes the field of reflection on rationality for shaping the scientific and technological future (Grunwald 2000a).

Rationality in the sense indicated, however, extends beyond purpose-related rationality, since it must also take into account the dimension of defining purposes and goals and, as such, also the acceptability of possible side effects, not just the selection of means to arrive at specifically defined

---

[4] Gethmann (1979) introduced the notion of a "pre-discursive agreement" serving as a basis to enable discoursive communication at all. A pre-deliberative agreement, in distinction to a pre-discoursive agreement, covers all implicit or explicit preconditions, predicisions and premises which serve as a basis for a *specific* deliberation, not for a discourse at all. Therefore, pre-deliberative agreements include more substantial elements than prediscoursive agreements (Grunwald 2000a).

ends (cf. Habermas 1987, Rescher 1988, "evaluative rationality"). This distinguishes the concept of rationality used here from a mere technological understanding. The rationality of decisions and actions in the sense of their validity "for everyone" requires a reference both to the appropriate selection of means for established ends as well as to the "reasonable" determination of ends. Purpose-related rational actions can be irrational, if their purposes are irrational.

4.3.2
Dimensions of rationality

The pragmatic analysis of the use of the concept of rationality and its fundamental structure leads to the following inherent features of rationality assessments (Grunwald 2000a):

Dimension of relationality
Attributions of rationality to actions or decisions are not independent from pre-decisions but are valid only relative to pre-defined criteria. At first, they are relative to a normative catalogue of criteria what is seen as rational. Such criteria are, for example, requirements of consistency and coherence or the claimed suitability of invested means/ends-relationships. Secondly, they are dependent on the pre-decision what shall be understood by the scope of the notion of "everyone" mentioned above. The scope of "everyone" may, in specific contexts, be restricted to a limited number of persons or to certain groups; in other cases the "everyone" may include the whole society or the entire humankind including future generations. The requirements for trans-subjectivity will differ, depending on the range of the claimed trans-subjective validity. And thirdly, rationality assessments will depend on the knowledge available at the time the assessments are performed: a specific decision may be assessed as rational relative to a specific amount of knowledge but may be less rational if new knowledge becomes available.

Dimension of procedurality
Understanding rationality as a "designation of the ability to develop procedures for discursive satisfaction of claims of validity, to follow them, and to control them" (Gethmann 1995, p 468; translation A.G.), opens up the path for the procedural operationalisation of rationality (instead of presupposing a *substantial* concept of rationality like the economic concept of homo oeconomicus). The reason-based recognition of validity claims (Habermas 1988, p 445) in discourses constitutes the procedural means of determining whether an action or decision complies with the rationality criterion of understandability "for everyone". Conflicts accompanying assessments of past actions for the purpose of learning, on the one hand, or the weighing ex ante of competing technology policy options, on the other, are pragmatic causes of deliberations including rationality assessments. This deliberative advisory model contains

both technology-related and practice-related rationality. It avoids both the descriptive (technocratically) narrow interpretation of older style technology impact assessments or of decision theory as well as the assumption of the inaccessibility of normative questions for rationality considerations. The claim is that it combines the concept of end/means rationality and ethical reason. Briefly speaking: rationality assessment is procedurally done by deliberative argumentation.

Reflective dimension
Rationality assessments of actions and decisions are reflective in the way that the conditions of validity of the knowledge and orientations used in these assessments are taken into account. Accordingly, the resulting rationality assessments are considered as *fallible*. The uncertainty and incompleteness of the knowledge used and the limitations of the knowledge on orientations involved are leading to an inherent risk of assessment (Grunwald 2000a). It belongs indispensably to rational assessment to include the consideration of its own risks and limitations into the advice given to decision-makers. If this were not the case the concept of rationality would be understood in a "rationalistic" way promising unfulfillable ideas of certainty guarantees. The reflective dimension is highly important for the decision-maker because he/she has the right to know about the risks included in the advices from TA. Furthermore, this dimension is inherently related to the possibilities for *learning*.

4.3.3
Practical rationality for shaping the future[5]

Technology policy decisions serve the purpose of prestructuring the future by promoting or selecting individual options or entire bundles of options out of the existing diversity of possibilities and in doing so steer technology development in certain directions or rule out certain other directions. By seeking to exert specific influence on the potential of future players to undertake actions, what is involved in connection with these policy activities is *planning in the general sense of an intended exertion of influence on future potentials for action.* What is involved here is neither the planning of technological products nor the planning of technology development as a whole but the rationally shaping of framework conditions for further technological development. The use of the concept of rationality (Rescher 1988) in planning and decision-making in dealing with problems of technology development (Grunwald 2000a/b/c) allows to conceptualise an intentional approach to the future ("shaping technology") in consideration of the impossibility to pre-determine the future by planning. Viewing the future from the standpoint of planning rationality must not lead to an assumption of the ability to create pre-figured

---

[5] This chapter summarises earlier work: Grunwald 1999, Grunwald 2000a/b.

and fixed future conditions. Rational planning, too, constitutes an activity *involving risk* (cp. the reflective dimension of rationality above).

Scientific uncertainty, the lack of verifiability of scientific knowledge in a definitive sense, the provisionality and incompleteness of scientific knowledge, the impossibility of giving scientific guarantees are widely acknowledged. Setting priorities in technology policy, definition of goals and measures, therefore, always require an element of *flexibility*. Because of the unavoidable hypothetical constituents of the basis for decision-making and action, due to the known uncertainty of knowledge about the consequences of implementing political measures, due to the possibility that new methods of action could become known, and due to the possibility of a change of goals in the non-scientific area (e. g., "change of values" in society), in view of the openness of the future in an open society, options which permit flexibility of action must be built into decisions which can have far-reaching consequences.

This means, in particular, that the rational shaping of technology development, such as in the field of research funding or government regulation, must always be temporary planning since by definition it is *flexible planning*. Shaping technology relative to criteria of planning rationality does not lead to predetermined final states but consists of acting under the obligation to *reflect* (the normative premises, the state of knowledge, the purposes and the interpretation of the relevant context aspects). Shaping the boundary conditions for technology development being the task of technology policy, the task of TA should consist of this permanent reflection under aspects of instrumental, evaluative and cognitive rationality (Rescher 1988). The quality and reason of flexible plans depends on the instrumental rationality in terms of means and ends as well as on the ethically justified choice of reasonable purposes, expectations and visions (Grunwald 2000c). This implies that *societal learning* at various levels becomes a decisive activity in shaping the future by the social enculturation of technology. Technology Assessment can be considered as an instrument to support societal learning in the field of technology development and technology policy by scientific means. Within "directed incrementalism" resulting from the considerations and reflections above (Grunwald 2000c), the direction of action and decision is maintained but not fixed. Permanent reflection on the goals and the means to attain them lead to incremental changes of direction in the development, of the goals as well as of the measures to reach the goals. This change, however, does not occur on the basis of chance events and does not show an erratic behaviour. This kind of development allows us to get closer to the envisaged area of goals and to take into account the short-ranged flexibility requirements (which may lead to incremental changes of direction). In this way, shaping the future is imagined as a stepwise approach including many small and reflected decisions against a background of rather stable normative orientations. This model allows a maximum of learning effects in approaching the future.

The possibility of learning and rationality assessments are interconnected in that only statements for which arguments can be given can be improved. If it were not possible to base actions on rational criteria, the only means of learning would be to use the (extremely inefficient) method of trial and error. Rational reflection *ex ante* thus the basis for reflection *ex post* on the reasons for success or failure and forms a necessary condition for learning (Habermas 1988, Grunwald 1999). If the action leads to success the assumptions involved are proven to be suitable. In the opposite case the site in the argumentation structure might be identified where a false or unsuitable assumption caused the failure. Knowledge improved in such a way could be used in the next decision-making situation involving this type of knowledge to improve the expectability of success by avoiding this type of failure. *Vice versa*, if one would, as an example, rely only on unfounded prophecies there could not be any chance for learning, neither in the case of success (which, indeed, might occur though its unfoundedness, see above) nor in the case of failure. There could not be any "lessons learnt" for the next case. In this way, such cognitive issues are important for learning. More generally speaking, learning consists of a special type of communication between various actors with regard to certain cognitive mechanisms. Ongoing observation, modelling, simulation, evaluation of measures for political management are just as necessary as openness for decisions modifiable by scientific and social *learning processes*. The task of rational TA is, therefore, to investigate and incorporate possibilities for learning and adaptation for making flexibility in the implementation of instruments of technology policy possible.

4.4
Quality requirements for interdisciplinary Technology Assessment

In a comprehensive concept of TA the question arises how these methodical ideas of problem-oriented research, rationality assessment, flexible planning and the enabling of social learning processes can be made to work in practice. Where are the organisational instruments to realise such far-ranging ideals? In this chapter a step towards an answer is offered by analysing the problem of safeguarding the quality of interdisciplinary research in general, applied to the requirements of rational TA. In the remainder of this paper it will be shown in which way the Europäische Akademie deals with these requirements.

In TA it is self-evident and has been exhaustively discussed that it is necessary to cross disciplinary borders, with all of the well-known methodical, communicative, and organisational problems this entails. Integrating disciplinary knowledge, defining interdisciplinary projects and talking across disciplinary borders are problems well-known in TA. In the following, a specific sub-problem is focussed on which has been discussed rather scarcely: how to safeguard the quality of integrative research in TA.

Integrative research is faced with large expectations concerning its problem-solving contributions. The results of TA are intended to influence social and, above all, political practice. However, there are difficult conceptual and methodical problems and challenges still unsolved. One of these problem areas is situated in the field of quality assessment. The fundamental question is how to distinguish between reliable results of TA, qualified to be embedded into political problem-solving strategies concerning technology, and nonsense, created by arbitrary or misleading premises or by the use of incompatible tools or terminologies.

It is imperative that TA results are reliable, and that their quality can be judged *ex ante*. Otherwise, a simple process according to the approach of trial and error would be carried out – which, in societal affairs, is extremely risky and inefficient. The question whether political measures should be taken depends decisively on the quality of these integratively-developed results. How can one distinguish good integrative research from that of lower quality? Is it at all possible to apply a methodological judgement (i. e., normatively), or is it necessary to wait until a consensus in the experts' discussion begins to develop (which is not at all certain, but, on the contrary, rather improbable)? In technology policy, furthermore, there is seldom enough time available to permit waiting for scientific consensus. In order to be able to investigate these questions in greater detail, we should recapitulate the standards for judging the quality of scientific research.

Quality assurance in scientific research is oriented on the criteria of validity for scientific propositions: the requirements of general, personally invariant validity (transsubjectivity) and intersubjective comprehensibility. In the normal process of scientific research, these criteria are submitted to quality tests on two different levels – (1) on that of method, and (2) on that of the organisation of research (Grunwald 2001). Wholesale transfer of both of these types of mechanism to interdisciplinary research is not possible, because there is no science of "interdisciplinarity", nor is such a (meta-)discipline, which could constitute a meta-paradigm in relation to the disciplines, possible. When the validity of scientific propositions made on the basis of interdisciplinary research is to be examined, it is at the outset completely unclear in which scientific terminology this should be done. According to which criteria should the quality of an interdisciplinary result Z, developed out of a statement A (valid relative to the disciplinary framework A') and out of a statement B (valid relative to the framework B') be judged, if we have to assume that A' and B' don't have anything to do with one another? What happens to the disciplinarily assured quality in interdisciplinary aggregations and integrations? In the following some specific challenges to quality assurance of rational TA as interdisciplinary research are analysed.

(1) Societal problems as the starting points for problem-oriented research have to be formulated explicitly by *ascribing words and applying terminologies*. Sensibilization of politicians and of the public is often attained

by means of cleverly chosen catchwords (examples are "Waldsterben", the climate catastrophe, the ozone hole). Characterisation of problems in this manner, can, of course, mobilise interest, but is not necessarily an appropriate point of departure for scientific analyses. It seems probable that these notions transport additional, evaluative and normative connotations. Such terms, therefore, don't merely describe but include (positive or negative) normative expectations about the future. Normative expectations or fears are transported within such descriptions. These examples show that choosing the terminology of description is an act carrying normative implications, hidden in metaphors, visions or cultural backgrounds: descriptions are more than mere representations. The claim of rationality means that such implications are uncovered in rational TA.

(2) Socially oriented formulations of topics are, as a rule, relatively diffuse in comparison with the requirements on scientific exactitude. By defining, stating, and operationalizing non-scientific life-world topics and problems more precisely, *classifications* commonly serve as a method of structuring. Classifications and the categorisation of problems have, on the one hand, grave consequences, but result neither unambiguously nor conclusively from the original formulation of the problem. In the syndrome approach to global change research, for example, the symptoms of non-sustainable development identified could be aggregated in a completely different manner to syndromes, the syndromes, in their turn, could be summarised in a completely different manner in groups. In all classifications and categorisations, there is leeway for reorganisation, which requires explicit reflection – on the one hand – of their usefulness for solving the problems at hand, – on the other, in view of their possible side effects and implications. In particular, one must be aware of the risk of excluding fundamental aspects due to the selectivity of classifications, and of the possible risk of an unintentional underestimation of essential aspects as a result of an implicit system of priorities inherent in the classification. Classifications are not value-neutral; values included have to be made transparent.

(3) In integrative TA research, it isn't a priori clear, how and according to which criteria the borders of the systems considered should be defined, because in this case, cognitive interests of various sorts have to be combined. To begin with, it is decisive that the constitution of the system-borders is carried out *only under the aspect of relevance*. The – for the solution of the problem – ex ante apparently optimal choice of system-limits must be made in the intent of maximizing the chances of success (this is obviously a decision under uncertainty). In this case, there are no scientific criteria, but means/end-argumentation is decisive. If, for example, in climatological research, anthropogenic influences are to be isolated, other system definitions would be necessary than in the case of an investigation of climatic change within geological time-spans. Below this level, the question of system demarcations in the disciplines involved poses itself. Here, other types of arguments

can play a role – for example, previous experience with systems of the type concerned, common disciplinary usage, etc. Possibly, the various disciplines co-operating in the investigation would operate with different system demarcations. This can't be "forbidden" normatively, with reference to standards of quality; one can, however, expect that *interdisciplinarily comparable* system definitions be employed. The criteria for compatibility may then not be disciplinary specifications, but have to be derivable out of the problem to be solved. A (methodological) quality management would have to act on this basis, and reflect the system-demarcations chosen and their pragmatic compatibility relative to the non-scientific criteria of problem orientation.

(4) In integrative TA research, the quality of the results gained from models depends essentially on the *pre-empirical conditions of validity* for modelling. As an example, the problem that focussing or limiting the model to the "relevant" interactions is necessary, in order to make the model concrete, should be inspected more closely. In such cases *decisions on relevance* have to be made relative to the determination of the aims pursued. *Essential* aspects have to be distinguished from *unimportant* ones. This differentiation is, like the demarcation of a system, *selective*. The success of integrative research depends decisively on the question whether these relevance appraisals (which obviously are made under conditions of uncertainty) "succeed". If a "false" decision is made here, the subsequent empirical investigations won't be able to compensate for this fault, regardless of their quality. Results of integrative research which are based on models can claim validity only relative to these preceding relevance decisions. Methodological quality management in this area must, therefore, on the one hand reflect the expediency of the choice of models for solving the problem in question, on the other hand (as above), carefully examine the pragmatic compatibility of the various models used. Both of these deliberations are carried out with regard to non-scientific problem definitions, awareness of social problems, and also non-scientific priorities, which set the framework for relevance decisions in model-building.

(5) Whenever knowledge from various disciplines has to be integrated in a problem-oriented manner, the constituent bits of knowledge which have to be integrated are, as a rule, available only in a form which makes comparison difficult and an immediate answer to the question of the result of this integration impossible. It can, therefore, occur that the contributions of the various disciplines (either intentionally or because of objective circumstances, such as a lack of time or money) have been prepared in varying depth of detail, or differ in other qualitative characteristics. A particular challenge in this sense lies in the connection between quantitative and qualitative modelling, and accordingly, at the interface between natural-scientific (quantitative) and social-scientific (often qualitative) results. By integration of the results, the question poses itself whether the quality of the final, comprehensive result – technically spoken – should orientate itself towards the lower end of the scale,

or towards the weakest contribution, or whether it can at all be aggregated out of the quality criteria of the individual contributions.

The orientation of TA with respect to the formulation of social problems implies that the criteria of scientific research are constituted extra-scientifically. It has been shown that integrative research refers to socially relevant non-scientific decisions in essential pre-empirical respects. The quality of the results to be expected then depends essentially on the "quality", i. e., the adequacy of these decisions as to the problem under discussion and the responsibilities for solving it. In this way, in addition to *internal* quality criteria for scientific work *external* criteria have to be applied. Integrative research doesn't raise less stringent, but rather *higher demands* on quality assurance: observance of the usual disciplinary standards of quality is necessary, but not sufficient to guarantee the quality of integrative research. Otherwise the interdisciplinary results might develop into "science light", with unclear quality criteria, and open for a methodical "rule of thumb". This challenge is an important field for interactive and interdisciplinary deliberation. Otherwise TA would get into danger to lose external quality. If such fundamental and pre-empirical ingredients of TA like terminology, system boundaries or assessments on relevance were biased this deficit could not be compensated by the consecutive modelling or research.

The considerations presented above direct our attention to two different aspects: (1) the importance of *deliberations on relevance* for the "construction" of systems and for modelling, and (2) the requirement of *pragmatic compatibility* on all of the levels discussed, in order to secure the quality of interdisciplinary research.

(1) Judgements of relevance are made in various stages of the process of research in systems analysis. The question is: which possible aspects of the investigation, which interactions, or which contents of the subject matter are relevant for a solution of the problem concerned, and which are not. Further, at as early a stage as setting up the interdisciplinary team, decisions have to be made as to which disciplines and subject areas presumably can make relevant contributions, and which could be omitted. Obviously, the success and the quality of the subsequent interdisciplinary research (measured on the chosen orientation of the problem) is decisively determined by this type of preliminary judgement. Setting the wrong premises on the level of relevance can scarcely be compensated at later stages, not even by the most excellent work. Quality management of interdisciplinary research, therefore, has to analyse critically the implicit and explicit relevance appraisals in this sort of process. The construction and justification of relevance-structures belongs to the essential pre-empirical responsibilities of interdisciplinary research.

(2) Pragmatic compatibility is a postulate of methodical rationality in interdisciplinary research. The choice of system demarcations, models, and methods in the various disciplines involved can not be made by these disciplines arbitrarily, nor exclusively from the perspective of the individual

discipline. Orientation on the problem in question is decisive for concrete demands on compatibility. No incompatibilities can be tolerated here, because the success of the project would otherwise be endangered; but strict theoretical demands for consistency in all respects, e. g., that the same theory must underlie all of the models used, are also unjustified. In order to guarantee the quality of problem-oriented research to be understood as making the promised contribution to the solution of the problem, scientific purism is by no means necessary. A certain measure of heterogeneity and inconsistency may be tolerable in individual cases – in any case, as long as success on the level of problem-orientation isn't endangered.

Both types of quality-assuring measures are directed by the specifications of the societally-defined problem. The criteria for relevance decisions as well as for compatibility requirements have to be derived from this problem by means of *pre-empirical deliberation* independent from disciplinary thinking and inner-disciplinary relevance assessments. The criteria for relevance decisions and compatibility requirements are normative; it is not possible to justify them inner-disciplinarily but the extra-scientific orientation to the intended contribution to problem-solving has to be taken into account. The justification of the choice of those criteria extends to social questions and requires under certain circumstances to invest political or ethical judgement. The societal orientation of TA is not only characterised by the *application* of scientific knowledge to societal problems, but extends into the criteria for ascertaining scientific quality. Such pre-empirical and normative judgements on relevances often are related to questions of the type: which society we *want* to live in, on the basis of which understanding of nature and the environment, and which conceptions of humanity are to be realised. Such questions therefore concern fundamental elements of society's self-concept and have the following characteristics in common with political conflicts: plurality of value systems concerned, lack of a concrete focus group, and interlacing of systems.

It has become clear that, in addition to an *internal* quality assurance grounded on inner-scientific methods, some kind of *external* quality assurance must be applied which would tackle the appropriateness of such pre-empirical conventions and decisions with respect to the problem to be solved. The question now is how this external quality can be ensured by means of research organisation. Are there organisation principles assuring a maximum of external quality? What is the maximum of external quality which can be constituted by science alone? Where are the limits of science in this field?

4.5
External quality assurance within interdisciplinary expert groups

Integrative science refers to a pre-defined societal problem. The requirement that this be a non-scientifically defined problem of social significance leaves the questions open: who defines the problem, which persons, groups, and social subsystems are involved, which interests do they pursue in the process, and how do perception and construction of the problem take place (or rather, should take place) – in particular, the question of science's role in this process of definition and construction. Examples for – at least – shared responsibility for the development of social problem-awareness are the definition of environmental standards, the climate problem (cf. the role played by the IPCC), the question of conserving biodiversity, and the question of the global water regime. Elements of social construction of research on these fields are the problem formulation, classifications used and, above all, the relevance of decisions mentioned above. In the following, the approach of the Europäische Akademie – the expert group approach – is taken as an example of how to arrive at maximum quality of interdisciplinary research by inner-scientific organisational means.

The realisation of Rational Technology Assessment in projects of the Europäische Akademie takes place in interdisciplinary expert groups which are organised in accordance to a strict procedure. In order to prepare this discussion process one starts with a pre-project in which the general complex of problems has to be pre-structured, the relevant scientific disciplines have to be identified and finally the experts have to be chosen.

The procedure itself is based on an intense discussion in which the experts should develop well justified recommendations to act from an interdisciplinary scientific perspective. This should be based on both high level disciplinary research and constructive cross connections between the disciplines, in order to transform the initial multidisciplinarity into interdisciplinary[6] results. This intention is evaluated in several evaluation loops in which external experts participate. The following chapters describe the pre-project, the project and the evaluation-loops in detail:

4.5.1
Pre-Project: Pre-structuring the problem field

Social, political or ecological problems appear in public debates with controversial positions. If the problems concern technology like e. g. climate change, nuclear power or biodiversity, a Rational Technology Assessment has first of all to pre-structure the ongoing discussion: What are the arguments? How are they covered by polemic statements? Is it possible to identify societal opponents? How are they organised? Reflecting on an ongoing discussion by answering questions like these leads in general to a reformulation of the social problem by simultaneously tracing the debate back to a more rational tier (cp. the above mentioned catchwords). This is where classification and categorisation of problems starts. The reformulated social, political, ecological

---

[6] For the differentiation between multidisciplinarity and interdisciplinarity cf. Mittelstraß (1998, p 32)

problem to be tackled by Rational Technology Assessment is the basis for the selection of the *relevant* scientific disciplines.

In general a single scientific discipline can be identified which is responsible for technical perspectives on the problem. Concerning nuclear power stations this would be physics and in the case of climate change it may be climatology. Economic aspects have to be considered, due to the fact that in most cases cost-benefit analysis are of interest. Moreover so called "reflecting sciences" like jurisprudence, philosophy and sociology could contribute to solve the initial problem and must be taken into consideration. As already mentioned, the selection of a particular scientific discipline should be justified relative to the problem to be solved. As a result of this justification process it becomes obvious that sub-disciplines of the well known university disciplines must be considered. For example one may need nuclear physicists and engineers of nuclear power stations to get technical details on safety aspects. Philosophy of Science as a (sub-)discipline of philosophy has a special role to play within Rational TA not only for disciplinary methodology but also for reasoning in connection with pragmatic compatibility. Moreover the justification process on relevant scientific disciplines will lead to several degrees of relevance. Some (sub-)disciplines may appear as to be absolutely necessary in any case, some others are just touched on special aspects of the problem.

All (sub-)disciplines found to be relevant should participate. But due to organisational reasons the group should be as small as possible, because a truly interdisciplinary discussion between experts with a different disciplinary background is very time-consuming and more so the bigger the corresponding group is. Therefore it is sensible to establish different degrees of participation according to the degrees of relevance of the scientific disciplines. The degrees of participation are for example "being member of the expert group" or "writing an additional study" or "join the few associate meetings in order to give evaluative remarks".

Up to this point it is assumed that the pre-project was organised by a project manager, typically a scientist with a relevant scientific training. He turns the initial social problem into a provisional work programme for an interdisciplinary expert group. For the final composition of the project group it is sensible to call in a *core group* of scientists out of the two or three disciplines found to be indispensable, e. g. the scientific discipline responsible for technical aspects and philosophy of science. The purpose of this core group is both the scrutinising and refining of the provisional work programme and, connected with that, the development of the final set of scientific disciplines, including the (relevance) decision in which manner the particular colleague will be participating. The decisive importance of this setting of the course has already been underlined above by pointing to the difficulty to correct "false" decisions.

Once a scientific discipline is determined to be relevant for the discussion, the problem arises which expert out of this discipline should be chosen. The following are some aspects concerning criteria for the recruiting of the experts.

- The experts should be selected referring to the real world problem to be solved. Due to the fact that just one or two experts from every (sub-)discipline can be invited to join the project group, all experts must be able to represent the major paradigms of his/her discipline, which is a task the experts have to agree to[7]. He also has to advocate for stakeholders connected to his discipline[8]. For this purpose a kind of overall view is needed.
- The experts must be interested in *interdisciplinary* discussions. This is meant in a procedural way, which involves, for example, a pre-discursive[9] and moreover the above mentioned pre-deliberative agreement. An important point, of course, is that all statements must be supported with comprehensible evidence and arguments. In an interdisciplinary debate this means that the arguments must be comprehensible to *all* participants[10]. Another point is that concrete recommendations for action should be given. Dissenting votes should be exceptional and have to be supported by strong arguments as well. Due to the fact that, in general, every scientist is an expert in just one scientific field, these rules presuppose a very intensive debate. Not every expert is prepared to participate in such a strenuous procedure.
- The selection procedure generally leads to several experts of whom one or two per discipline or sub-discipline will be appointed to participate in the project group. Having equal and small numbers of experts out of every scientific community is not only sensible for organisational reasons, but also helps to *avoid dominance* of particular disciplinary views. Every expert has to convince the experts of other disciplines of his argument. Hence, in the ideal case, the resulting memorandum will not be biased by the aims of one specific discipline or particular interests, it will be scientifically balanced.

At the end of the pre-project a group of experts has been appointed in which all scientists out of the disciplines found to be relevant are selected on

---

[7] However, there is always some doubt left, as to whether experts will defend paradigms of others with the same accuracy as their own ones. Therefore other experts are invited to forward their arguments at supplementary meetings (Kickoff, Midterm, etc.; see below).

[8] Here, this advocacy is not claimed to be complete. Nevertheless an expert of medicine for example has at least an idea about the stakes of the patients.

[9] Cf. Gethmann (1982)

[10] Cf. "justifiable to anyone else" (Rescher 1988; Gethmann 1995) or "practical reasoning" (Lorenzen a. Schwemmer 1973, Lorenzen 1987, Gethmann 1995).

the basis of rational arguments and a work programme has been worked out, scrutinised by these experts.

Example: One project of the Europäische Akademie investigates the question: In which areas is it sensible to replace capabilities of human beings by actions of autonomous robots?

The scientific discipline responsible for technical aspects would be robotics. Cost-benefit analysis has to be taken into account, because it is important to know if a robot acts cheaper or not. The core group would consist of an expert out of robotics and an experts out of economics. In addition to that moral aspects have to be taken into account in order to check if there are some areas in which we do not *want* to have a robot acting instead of a human being. One also needs an expert out of jurisprudence to answer questions concerning the liability of autonomous robots. Who is responsible for damages caused by a robot?

If one goes into detail it appears that several sub-disciplines must be considered. An expert for artificial intelligence is necessary who covers aspects concerning software, as long as the expert for robotics focuses on hardware. Therefore he/she should participate as a full member of the project group. Aspects concerning the direct contact between robots and the nerve system of human beings, which is important in connection with prostheses, are not in the main focus of the project, therefore it was decided to invite an expert out of this field to the (one or two) meetings of the project group in which this topic will be discussed.

The selection of the experts out of the scientific disciplines found to be relevant takes place relative to the initial problem, so as to it is sensible to chose an expert out of jurisprudence, who has at least some ideas about computer science as well. But this expert explained that he would not be able to prepare the comparison between different national laws in Europe. Therefore an additional expert has been asked to write an additional study about this law comparison.

(The complete pre-project of the robotics project is described in Decker (1997))

4.5.2
Project

The following structure outlines the procedure to organise an expert group in order to reach an interdisciplinary discussion based on rationality (Gethmann 1979; Gutmann and Hanekamp 1998; Rescher 1988):

- Introductory Phase
  Every expert is an expert just in his discipline. Therefore, an interdisciplinary debate starts with disciplinary statements in which the experts present their perspective on the initial problem. These statements are the

basis for the following discussion about concepts, their definitions and underlying assumptions, which are common in particular disciplines and consequently rarely discussed. In this way, underlying (disciplinary) assumptions are made explicit. At the end of the introductory phase the core notions to be used during the project should be defined in order to reach a common *terminology*.

In this phase first proposals are made on disciplinary system demarcations and on the disciplinary methods to tackle the problem. The structuring process started in the pre-project has to be finished and the classifications and problem categorisations made so far have to be scrutinised.

- Analytical Phase

  After this process synthesising the perspectives of the different disciplines is approached by reflecting and incorporating the particular others. In this discussion, the development of transparent argumentation chains starts, where the arguments come from the different disciplines and have to be accepted by the other disciplines. This results in one or more chains of argument supported by all participants. One focuses on combining the perspectives in a rational manner by deciding about the relevance of different arguments. In addition to that the compatibility of the disciplinary perspectives must be scrutinised. That is where the pragmatic compatibility between the disciplines has to be developed. Considering the initial problem as a basis, the tolerable degree of heterogeneity and inconsistency has to be negotiated referring to argumentation.

- recommendation phase

  In the last phase, these argumentation chains have to be extended into the area of pragmatic consequences, i. e., the initial impulse of the TA-project. Again this should be done by reference to rational arguments. In this constructive phase, the experts have to decide on recommendations for actions to be taken. In the procedure described so far, this involves selecting one argumentation chain out of all of those developed so far.

  To choose one of the argumentation chains is not evident *per se* because other TA-approaches take it for sufficient to present the alternative solutions without any further recommendations in order to leave the choice to the decision makers. But even well justified seemingly pure descriptive forecasts and foresights are carrying normative implications. Therefore it is just consequent to choose one recommendation to act, justified by transparent reasoning, because it makes the normative character of the results obvious. The recommendations to act should be formulated in the sense of the above mentioned flexible planning and directed incrementalism. This means 'virulent' stages of technical development can be defined as *milestones*. When they have been reached, one has to decide about future alternatives or one has to reflect on the development reached so far. Not only technical milestones are sensible. It is also possible to define economical milestones when very expensive developments are on the agenda or societal milestones which entail a public opinion poll to get an idea about change of values.

The procedure of the interdisciplinary discussion once started should not be taken as not variable. As already mentioned the structuring of the problem field, including several steps of classification and categorisation of problems, has to be finished during the project phase itself. Therefore it may be possible that the relevance of a particular scientific discipline has been misjudged. On the other hand one has to keep in mind the crucial importance of this setting of premises and initial relevance decisions. That's why it should be possible during the project phase to call in additional experts or to ask for completing studies whenever they appear to be relevant. The risk of excluding fundamental aspects should be minimised in that way.

Referring to the example again one recommendation to act could relate to the classification of autonomous robots, which is an important point from a legal point of view. Up to now they are nothing else but automatic machines and it is not necessary to define new rules for robots. But when they are able to learn specific tasks which could be defined as a technological milestone, one has to think about this classification again. It may be sensible to take them, from a legal perspective, as animals.

4.5.3
Evaluation Loops

Evaluation processes within a scientific discipline are well established. In interdisciplinary research the high quality of the disciplinary contributions is necessary and must therefore be evaluated. But in addition to that the composition of the disciplinary perspectives, the transparency of argumentation, the justification of relevance decisions must be scrutinised as well.

Due to the fact that experts "for the future" do not exist, the question arises how interdisciplinary research should be evaluated. In the following several interdisciplinary evaluation loops will be presented, which have to pay attention to both, the disciplinary contributions and the interdisciplinary composition of reasoning. This is based on the assumption that optimal transparency and conclusive justification can be reached by continuous evaluation by external scientific experts. Controversial disciplinary perspectives, which may have been made out, should explicitly be taken into account during the evaluation processes. The discussions on these controversies are important for the evaluation of disciplinary quality aspects.

The crucial criterion for the evaluation of the interdisciplinary quality of the research is the initial social, political or ecological problem. Quality control of interdisciplinary research without reference to a "real world" context remains arbitrary. All relevance decisions are of special interest in this context, because by means of them it is justified what is in the focus of the project and what is not. Especially the potential losses caused by these relevance decisions must be analysed. Referring to the reflective dimension of rationality, the project group itself has to be aware of the uncertainty and incompleteness of knowledge in order to estimate the inherent risk of the assessment. But, however, this self-reflection has to be complemented by external evaluation, because as important as the resulting recommendations to act are the considerations about risk and limitations of the assessment.

Referring to the remarks on quality control of interdisciplinary processes, i. e. judgements of relevance decisions and the pragmatic compatibility, it becomes obvious that the evaluation process itself must be an interdisciplinary endeavour. As well as the arguments for these decisions have to be negotiated in the project group, the evaluation can only take place in interdisciplinary discussion by transparent reasoning. The evaluation processes described below are always interdisciplinary discussion processes.

Pre-project/Work programme
The provisional work programme contains the transformation of the initial social, political or ecological question into a scientific, multidisciplinary question and is therefore essential for the success of the project. Remembering the above mentioned non-scientific criteria involved in this process and the means/end argumentation to be considered, the necessity for evaluation is obvious. The work programme has to pass two evaluation processes. Firstly an internal process by the two or three experts out of those scientific disciplines found to be absolutely necessary. That is why they have been called into the "core group". They focus on the relevance of the different scientific disciplines and the degree of participation in the project and reformulate by that the work programme. That's where the course is set for the whole project, therefore a second evaluation is made by external scientific experts who are asked to scrutinise both the work programme and the justification of the participating disciplines.

As mentioned above, the pre-project is a selection process. Several premises have to be set and justified referring to the initial societal problem. Scientific classifications are not value-neutral and assessing relevance is an endeavour with normative implications. The justification of these relevance decisions must be evaluated.

Start of the project/Kick-off Meeting
The experts participating in the project group are asked to represent the major paradigms in their (sub-)disciplines and have to take that into account in their initial statements. But, however, every expert has his personal perspective on his scientific discipline. Therefore it is sensible to ask other experts from the same discipline to give additional statements. If it was possible during the pre-project to identify conflicting disciplinary perspectives, one would make sure that these conflicts are presented at the beginning of the project. This is helpful for the respective colleague who gets a reflection on his own perspective and for the other participants of the group who get a second

expert opinion. The hearing of these external experts has to be realised in a meeting[11] (Kick-off-Meeting), because it should broaden the initial discussion basis. Therefore it is necessary that the disciplinary statements are presented not only to the members of the project group, but also to the external experts of the other disciplines. This is of crucial importance for the optimal choice of system limits, which has to be made under the intent of maximising the chances of success. A second opinion on these limits, given by the external experts invited to the Kick-off-Meeting, allows relative argumentation.

The Kick-off-Meeting is an evaluation process in the sense that external experts give their statements on record and these statements have to be considered by the project group. Either by taking them into the argumentation or by well justified non consideration.

Work in progress/Midterm-Meeting
Work in progress should be evaluated for two different purposes. Firstly, the result reached so far must be scrutinised. How did the experts define and justify "their" system limits? How successful was the combination of the disciplinary perspectives? Was it possible to reach a pragmatic compatibility between the disciplinary systems? Was it possible to present transparent chains of arguments? Secondly, new aspects, up to now not considered, could be given to record. Again these inputs must be taken into consideration by the project group anyway, by accepting or by justified refusal.

The Midterm-Meeting should be organised to realise both the evaluation of disciplinary results by inviting the relevant external experts and the assessment of the interdisciplinary results by giving them to all participants of the Midterm-Meeting. Especially for the latter a briefing of the experts will be necessary. Without having an idea about the initial societal problem, about the transformation processes during the starting phase of the project, about means to be chosen and ends to be reached, a serious evaluation would not be possible. Therefore the preparation time for an interdisciplinary evaluation is clearly longer than for pure disciplinary evaluations.

Final Results
In the last phase of the project the experts will develop normative criteria in order to choose one of the argumentation chains worked out so far. By formulating the concrete recommendations to act the memorandum itself will be finished. This is where the 'implicit normative aspects' (see above) become explicit by transparent reasoning. It is of crucial importance that this step as well is scrutinised by an interdisciplinary expert group. The evaluation process should focus on the justification of relevance decisions, on the compatibility of the interdisciplinary solutions, on the transparent presentation of the results, and on the adequate formulation of the milestones.

---

[11] Instead of inviting the experts out of the relevant disciplines one after another.

As already mentioned for the Midterm-Meeting this evaluation process must cover the whole project and will therefore need time.

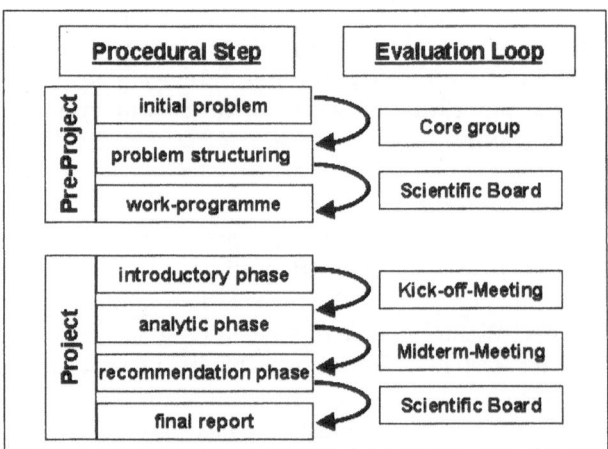

4.6
Discussion

4.6.1
The interdisciplinary procedure

The project described so far consists of an intensive interdisciplinary discussion accompanied by several evaluation processes. If one takes into account the preparation time for the Kick-off- and Midterm-Meeting, two years seems to be a realistic duration for the whole project. If we take additional six months for the pre-project we reach nearly three years for the whole undertaking. In order to realize the deep cross-correlation between the disciplines to turn multidisciplinarity into interdisciplinarity, the project group should meet monthly. Where are the chances and where are the limits of such large-scale interdisciplinary research?

"Interdisciplinarity on all levels" could be the headline for the procedure presented so far. But, however, it should not be forgotten that the *high quality of the disciplinary research* is indicated as to be absolutely necessary. Therefore additional experts are called in to the Kick-off- and Midterm-Meeting to discuss the disciplinary statements within a discipline. During the whole project at least two or three experts from the disciplines found to be relevant should participate in one of the steps sketched so far. The discipline-intern evaluation is of crucial importance especially by defining the disciplinary system limits. One could argue that the optimal disciplinary expert opinion could only be reached by intense discussion of all experts out of this discipline, which may be true. We propose to take two or three experts from the same discipline and let them discuss in the presence of scientific experts from other disciplines. The latter can be seen as a kind of jury, like explained below. When controversial disciplinary perspectives appear, the project coordinator should take care that the opposed view of the expert, who is member of the project group, is taken by an expert invited to the evaluation meetings.

A main criticism against expertocracy is that it is hardly avoidable that experts express their political orientation 'covered' in their scientific statements. Van den Daele and Döbert (1995) report that 'political orientations could be neutralised by other competent conversation partners in intensive discourses' (translation MD). This takes place by identifying and uncovering these biased inputs in conversation. The same happens to contradictory arguments of scientists from the same discipline, the so called 'expert dilemma' (Nennen and Garbe 1996). The other participants of the discussions are able to evaluate two different argumentation chains when they are presented in a discussion. The well known rhetoric tools like telling only half of the truth, using catchwords and phrases instead of arguments, interpreting empirical data in just one direction, and so on, can be uncovered by discussion. Van den Daele and Döbert are talking about 'abolition of selectivity' (translation MD) and report a kind of 'intellectual cleansing process' (translation MD) by discussions between experts and laypersons. In our case the role of the "jury" is taken by the experts out of the other disciplines, one would talk about a scientific self-cleaning process which may lead to well-balanced interdisciplinary results.

Due to the fact that each participant is an expert only in his own discipline, all disciplinary statements have to be explained to 'laypersons' with scientific background. Scientific background in this context can be described as "being familiar with formulating and understanding scientific reasoning". Therefore the disciplinary statements will have a higher level of transparency concerning the choice of methods than in discussions confined within single disciplines. High transparency is also necessary for the definition of the disciplinary system limits, because that is where the possible links to other disciplines have to be presented.

The societal learning process as described in paragraph 2 seems to be prepared: The process described so far is a kind of *ex ante* reflection and the whole procedure including the evaluation-loops is designed for transparent argumentation up to concrete recommendations. The latter are connected with the concept of flexible planning. The success of the project concerning the method of accompanying such development by reflection is to be achieved by establishing milestones, which have to be suitably formulated.

4.6.2
Trans-disciplinary quality of TA?

Up to here, the discussion of this contribution focused on the disciplinary and interdisciplinary procedures. Concerning the recommendations to act and the milestones, the border between science and society has been crossed and we have to discuss transdisciplinary and normative aspects. At the end of chapter 4.3 it was explained that the criteria for the scientific TA-research are extra-scientifically constituted. This concerns especially the normative and pre-empirical judgements and the questions connected with them.

The project as well as the evaluation-loops described above remain in scientific contexts. It was stressed that disciplinary experts are also competent laypersons and of course members of society. But the question arises whether they should be competent in questions like: In which society do we want to live? Which concepts of humanity are to be realised? Why should they be able to represent the plurality of value systems?

The evaluating discussions are described as needing intensive preparation in the sense that the evaluators should bear in mind the initial social, political or ecological problem. They are also informed about the transformation processes which led to the work programme and they are asked to decide about the adequacy of the recommendations. This includes for example the reflections on expediency of the choice of models or the compatibility between these models, both aspects which can only be assessed by referring to the original problem (see chapter 4.3). However, no claim can be made to any kind of completeness concerning the extra-scientific evaluation process. Of course the recommendations to act are aimed at making a proposal to solve the original problem. The evaluation processes, however, are mainly interdisciplinary albeit scientific in their structure.

According to that the question arises as to what an optimal extra-scientific evaluation should look like. What changes, when NGO's and politicians[12] are invited to the evaluation processes? The discussion basis would be broadened by these additional perspectives. And certainly the resulting recommendations to act would be changed, due to these additional inputs. However, from our point of view, it seems to be sensible to discuss the results of the scientific Rational TA afterwards with NGO's and politicians, due to the societal learning processes. In an *ex post* analysis it may be of interest to scrutinise the results 'step by step'.

One argument for the participation of politicians and NGO's may be that the recommendations to act would be prepared for direct transformation into public policy. From our point of view, Rational TA should be seen as a scientific consulting (providing knowledge as a basis for action) and therefore the separation from political decisions is important for us. Of course, the

---

[12] This enumeration could be extended with stakeholders, citizens, laypersons. Some ideas concerning Rational Technology Assessment and "expertise of laypersons" can be found in Decker, Neumann-Held (2001).

arguments of the consultation are legitimised by quality-controlled interdisciplinary research, and therefore not easy to overrule in a democratically political discussion. Political conflicts should be resolved by political procedures. To resolve political conflicts concerning future technology it is decisive to have scientific consultation – among others!

References

Bechmann G, Frederichs G (1996) Problemorientierte Forschung: Zwischen Politik und Wissenschaft. In: Bechmann G (ed) Praxisfelder der Technikfolgenforschung. Konzepte, Methoden, Optionen. Frankfurt, p 11–37
Brown HI (1988) Rationality. London/New York
Decker M (1997) Perspektiven der Robotik. Überlegungen zur Ersetzbarkeit des Menschen. Graue Reihe Nr. 8, ISSN 1435-487X
Decker M (2000) Replacing Human Beings by Robots. How to Tackle that Perspective by Technology Assessment? In: Grin/Grunwald 2000, p 149–168
Decker M, Neumann-Held E (2001) Between Expert TA and Expert Dilemma. A plea for Expertise in Technology Assessment. To be published
Funtowitz S, Ravetz J (1993) The Emergence of Post-Normal Science. In: R von Schomberg (ed): Science, Politics and Morality. Kluwer Academic Publisher, London
Gethmann CF (1979) Protologik. Untersuchungen zur formalen Pragmatik von Begründungsdiskursen. Suhrkamp, Frankfurt
Gethmann CF (1982) Protoethik. Untersuchungen zur formalen Pragmatik von Rechtfertigungsdiskursen. In: Elwein Th, Stachowiak H (eds) Bedürfnisse, Werte und Normen im Wandel, Vol 1. München, pp 113–143
Gethmann CF (1995) Rationalität. Enzyklopädie Philosophie und Wissenschaftstheorie. J. Mittelstraß (ed). Stuttgart, p 468–481
Grin J, Grunwald A (2000, eds): Vision assessment: shaping technology in 21st century society. Towards a repertoire for Technology Assessment. Heidelberg et al.
Grunwald A (1999, ed) Rationale Technikfolgenbeurteilung. Konzeption und methodische Grundlagen. Berlin Heidelberg New York
Grunwald A (2000a) Technik für die Gesellschaft von morgen. Möglichkeiten und Grenzen gesellschaftlicher Technikgestaltung. Frankfurt
Grunwald A (2000b) Technology Policy between Long-Term Planning Requirements and Short-Ranged Acceptance Problems. New Challenges for Technology Assessment. In Grin/Grunwald 2000, p 99–148
Grunwald A (2000c) Handeln und Planen. Philosophische Planungstheorie als handlungstheoretische Rekonstruktion. München
Grunwald A (2001) Integrative Forschung zum Globalen Wandel: Herausforderungen und Probleme. In: R. Coenen (ed): Integrative Forschung zum Globalen Wandel. Frankfurt, p 19–41
Grunwald A, Saupe S (1999, eds) Ethik in der Technikgestaltung. Praktische Relevanz und Legitimation. Heidelberg Berlin New York
Gutmann M, Hanekamp G (1998) Wissenschaftstheoretische Grundlagen Rationaler Technikfolgenbeurteilung. In: Grunwald A (Hrsg) Rationale Technikfolgenbeurteilung. Konzeption und methodische Grundlagen. Springer, Berlin, Heidelberg, S 55–91

Habermas J (1987) Zwecktätigkeit und Verständigung. Ein pragmatischer Begriff der Rationalität. In: Stachowiak H (ed) Pragmatik. Handbuch pragmatischen Denkens. Hamburg, p 32–59

Habermas J (1988) Theorie des kommunikativen Handelns. Frankfurt

Janich P (1992) Grenzen der Naturwissenschaft. Erkennen als Handeln. München

Jaeger J, Scheringer M (1998) Transdisziplinarität: Problemorientierung ohne Methodenzwang. GAIA 7, p 10–25

Lorenzen P (1987) Lehrbuch der konstruktiven Wissenschaftstheorie. Bibliograpgisches Institut Mannheim

Lorenzen P, Schwemmer O (1973) Konstruktive Logik, Ethik und Wissenschaftstheorie. Bibliograpgisches Institut Mannheim

Luhmann N (1990) Die Wissenschaft der Gesellschaft. Frankfurt

Mittelstraß J (1998) Interdisziplinarität oder Transdisziplinarität? In: Mittelstraß J (ed): Die Häuser des Wissens. Frankfurt, p 29–48

Nennen H-U, Garbe D (1996) Das Expertendilemma. Zur Rolle wissenschaftlicher Gutachter in der öffentlichen Meinungsbildung. Berlin et al.

Paschen H (1986) Technology Assessment – Ein strategisches Rahmenkonzept für die Bewertung von Technologien. In: Dierkes M, Petermann T, Thienen V. v. (eds): Technik und Parlament. Berlin, p 21–46

Rescher N (1988) Rationality. Oxford

Van den Daele W, Döbert R (1995) Veränderungen der äußeren Natur partizipative Technikfolgenabschätzung. In: Funkkolleg Technik, Studienbrief 4. Belz Verlag, Hemsbach

Zweck A (1993) Technikfolgenabschätzung als gesellschaftliches Vermittlungsinstrument. Opladen

5 To assess rationality before anything else. A remark on the legitimacy of Rational Technology Assessment

Rob P. B. Reuzel

Abstract

This involves a comment on the chapter by Michael Decker and Armin Grunwald in this volume, that is, on what these author's have coined "Rational Technology Assessment" (RTA). And although the richness of Decker and Grunwald's account deserves a lot more, I will confine myself to only one aspect: the "rationality" of RTA. I believe that the kind of rationality exploited in RTA makes Technology Assessment expedient, rather than legitimate. To develop my argument, I take two steps. First, I give a brief sketch of the history of Technology Assessment until the rise of social constructivist approaches, and try to explicate the rationale of RTA on the basis of this sketch. In fact, I regard RTA as a new and courageous attempt to complement social constructivist philosophy with a feasible plan of work. Second, however, I argue that RTA simultaneously divagates from this philosophy, in that it neglects its legitimate basis by depending on expert knowledge too heavily. RTA resembles what Habermas has called a "technocratic model".

5.1
A brief history of Technology Assessment

Technology Assessment began to develop in the second half of the twentieth century. It was born as a cultural criticism and an early warning system. Typically, Technology Assessment in its early stages beared only a weak relation to technology policy (Smits a. Leyten 1991, p 1). It was to signal the unintended and negative side effects of using technology. The perceived fast and autonomous developments of technology were experienced as something scaring by large parts of the population. Mankind faced huge problems in the fields of environmental issues, medicine, and military. Especially in the west, people felt overcome by technology, and strived to retain their top position in the order of things. "TA should be viewed," Carpenter has said, "as a mild form of disenchantment with twentieth-century technology as it has been practised by the industrial nations of the world." (Carpenter 1983, p 118)

In the United States, the Office for Technology Assessment was founded in 1972. By then, Technology Assessment had become a kind of *research*. And what is more, it had become a kind of *policy* research. Hence, Banta and Luce could define Technology Assessment as "a form of policy research that examines short- and long-term social consequences (for example, societal, economic, ethical, legal) of the application of technology. The goal of Technology Assessment is to provide policy-makers with information on policy alternatives." (Banta a. Luce 1993, p 61) Many more such definitions exist.

Knowledge is power. The rationale behind Technology Assessment was, and still is, control on the basis of knowledge. Whether it is to prevent technology from having unintended side effects, or to confine expenditures, it is asserted that technology development must be controlled. Therefore, Technology Assessment is to provide information that enables governing the various aspects of technology development. In early phases of Technology Assessment, this involved anticipating future developments, and demarcating the field where these developments could have their impetus. Later, when Technology Assessment became institutionalised, it helped to confirm the grip on technology development in two ways. One, it served as a kind of policy research to support decision-making directly, that is, not conceptually, but instrumentally. And two, it served to verify the impact of policy already being pursued. I am being careful here, and do not wish to suggest, as Rapp does, that Technology Assessment opportunistically "can be exploited to *justify political aims*." (Rapp 1983, p 143; my emphasis) The task of verifying the impact of policy, of course, does not imply that Technology Assessment did no longer have anything to do with anticipation. However, fit in the framework of policy-making, its emphasis shifted from the theory of socio-technological dynamics towards providing policy-makers with manageable data on the impact of technologies on their social environment.

Attempts to control should not only be understood as conservative approaches, as Technology Assessment can, and in fact was, used to enhance technological and economic competence, too. Rip, Misa and Schot have observed that "our society has relied on a two-track approach that separates promotional activities from control and regulation. Our institutions are set up in this way, with regulatory agencies separate from technology-promotional agencies, and the approach is embedded in our culture." (Rip et al. 1995, p 2) It is a response to the dilemma of simultaneously being addicted to technology and confronted with the limits of what the world can bear. The rhetoric of Gago, at the time of his writing Portugese Minister of Science and Technology, is characteristic of the attitude policy-makers adopt towards Technology Assessment and social sciences in general:

It is only to the extent that Europe's policy-makers are provided with the leading edge of social science that they will be able to avoid the recourse to technocratic and bureaucratic solutions for the complex problems and be able to predict and forecast problems sufficiently in advance of catastrophes. Social sciences can enhance the social and economic potential of new knowledge within a new knowledge-based society and point out our continuing ignorance when it comes to the persistent problems of industrial society. The social sciences should be able to deal with complex societal processes involved in innovation processes and facilitate the application of innovation to the resolution of old problems. (Gago 1998, p 36)

However, Rip, Misa and Schot also suggest that the two-track approach has failed, mainly because it has proven to be difficult to control technology when it is already in its final stage of development. If Technology Assessment runs after the facts, the potential of Technology Assessment as a key component of attempts to control technological development is very limited. Obviously, the problems faced are too big to be countered by (public) awareness alone. In the eighties, it was recognised that the two-track approach is ideally replaced by a constructive approach, in which development and control are part of the same process. Technology development was to have a reflexive character and be based on early stage assessment of a broad range of aspects to *prevent* technology from having negative side effects.

Thus, the image of technology development as an autonomous wheel was refined, and subsequently abandoned. It was acknowledged that technology was as much the product of societal developments as it affected society. Technology was now seen as a societal aspect itself, the relation between technology and society being phrased in terms of interaction and mutual influence. "Social scientists have tended to concentrate on the 'effects' of technology, and on the 'impact' of technological change on society," MacKenzie and Wajcman wrote. "This is a perfectly valid concern, but it leaves a prior, and perhaps more important, question unasked and therefore unanswered. What has shaped the technology that is having 'effects'? What has caused and is causing the technological changes whose 'impact' we are experiencing?" (MacKenzie a. Wajcman 1985, p 2) Accordingly, Technology Assessment was redefined as the study of socio-technological interaction. This has come to be known as social constructivism. TA that is based on this philosophy is called "constructive Technology Assessment" (CTA).

CTA, according to Schot, "is based on the idea that during the course of technological development choices are constantly being made about the form, the function and the use of that technology, and consequently that technological development can be steered to a certain extent." (Schot 1992, p 37) It "is built around the attempt to anticipate effects or impacts of new technologies or new projects with a strong technological component. This is a core component of any TA effort. In traditional TA, the technology or the project is taken as given, and thus seen as a static entity. For CTA, the dynamics of the process are central, and impacts are viewed as being built up, and co-produced, during the process of technical change." (Schot a. Rip 1997, p 257) In this sense, it opposes the "positivist sciences," which, on a practical level, mainly involve hypothesising, measuring, and falsifying, and, on the level of methodology, the validity of scientific methods and claims. But most importantly, CTA reflects our creating our environment by use of technology. Constructivism, then, reflects a particular role for social sciences, including Technology Assessment, in society. Social sciences are regarded as an aid in construing alternatives for a better society, which is based on insight into the societal embedding of technology.

Nevertheless, it is difficult to grasp what all this amounts to. Although to recognise that technology is a social construct leaves the impression that technology development is in our hands and can be steered, it is not at all clear *how* technological change could be (re)directed to a desirable path. In its early phases, constructive Technology Assessment stood for a kind of enlightenment with respect to the interdependency of technology and society, rather than a set of tools that could readily be used to deal with it. Still, except for a few methods just mentioned, no definite answer matches the question how to do a constructive Technology Assessment. It is therefore that attempts to realise a workable programme for Technology Assessment are important. In this volume, Michael Decker and Armin Grunwald have sketched the outlines of such a practical approach, which they have baptised as "Rational Technology Assessment" (RTA).

RTA, in my opinion, is a typical product of the history of Technology Assessment briefly sketched above: the time has come for such approaches. It departs from several assumptions that may be called "constructivist". First, technology and society are viewed as intertwined. Therefore, the relation between technology in society is complex, and warrants a multidisciplinary approach. Partly as a result of the complexity and unpredictability of the technology-society relation, Technology Assessment involves dealing with un-certainty or risk. This implies the need to take small steps in technology de-velopment, and be prepared to reconsider goals and means at all times during the process. Furthermore, it is better to develop technology that is accept-able, than with hindsight judge technology that is already in an advanced stage of development. The usefulness of TA as a instrument for steering re-quires an integrative approach; therefore, multidisciplinarity should become interdisciplinarity.

5.2
Rational Technology Assessment and the assumption behind "rationality"

For details, I refer to the chapter by Decker and Grunwald in this volume. In this chapter, I wish to draw attention to one assumption behind RTA: the fail-ure of the contemporary political system to face the problems of technology in society adequately. Of course, from a macro-perspective TA in western societies *is* multidisciplinary in the sense that many different persons, each with his or her own expertise, assess technology. Furthermore, their accounts *are* integrated, that is, synthesised at a political level, although perhaps rather implicitly. What is wrong with this system?

If I understand correctly, advocates of RTA answer that the system lacks rationality. I take this to mean that it suffers from such noise as political intuition, economic interests and public emotions, causing short term acceptance to prevail over long term acceptability. As a result, it is inefficient, too. In short, it is *rather* political. In any case, it is unlikely to lead technocratic societies in a direction that makes them sustainable in the long run. We *need* (more) rationality, advocates of RTA claim. We need to know what indisputably *is* best for us, should we want to prevent our backyard from suffering the disasters that are already visible at the horizon. We need those who can see the rainforest in our backyard, and look beyond, that is, experts.

RTA, therefore, perfectly fits an "expertocracy". The term "rationality" alludes to expert knowledge and scientific standards of validity. (It is no coincidence that validity is heavily emphasised in RTA.) Even moral disputes are to be resolved within expert groups. Not only do experts have both deeper insights and broader visions, especially when they co-operate in interdisciplinary teams, they are also rational in the sense that they have the capability to refrain from personal emotions and preferences, and retrieve knowledge that is scientific, that is, beyond dispute. In fact, with RTA the need for a political system has been eradicated completely. Or has it? RTA might be expedient, rather than justified.

It is here, that Decker and Grunwald take a principally different position compared to advocates of CTA. In CTA the interdependency of technology and society is not only acknowledged, but also seen as an opportunity: technology development *can* be steered *through* its societal embedding. Currently, this has lead to an increased interest in stakeholder empowerment and participatory research. (Examples of these are included in this volume.) In RTA, however, society and the public as active players in technology development are distrusted. The interdependency of technology and society is recognised, but only as an object of study. In RTA, society is isolated in favour of the expert world.

5.3
Habermas on decisionistic, technocratic and pragmatistic models

Habermas distinguishes between three models of the relation between science (in general) and politics (Habermas 1996). The first is a so-called "decisionistic model", in which, in accordance with the works of Weber, scientific expertise and political practice are strictly separated. In this model politics is informed by scientific expertise, but is furthermore based on choice behaviour that escapes scientific modes of reasoning. In the "technocratic model", on the other hand, this state of affairs is entirely reversed. That is, scientific experts are dominant, not only when it comes to research issues, such as, for example, validity, but also with respect to the implications and use of scientific research. In this model, apart from the information used in decision-making, political procedures *as such* are based on scientific rigour. The politician, being not only informed, but also compelled by experts, is deprived of effective power of his or her own. Now, in the "pragmatistic model"

> the strict separation between the function of the expert and the politician is replaced by a critical interaction. This interaction not only strips the ideologically supported exercise of power of an unreliable basis of legitimation but makes it accessible *as a whole* to scientifically informed discussion, thereby substantially changing it. Despite the technocratic view, experts have not become sovereign over politicians subjected to the demands of the facts and left with a purely fictitious power of decision. Nor, despite the implications of the decisionistic model, does the politician retain a preserve outside of the necessarily rationalized areas of practice in which practical problems are decided upon as ever by acts of the will. Rather, reciprocal communication seems possible and necessary, through which scientific experts advise the decision-makers and politicians consult scientists in accordance with practical needs. (Habermas 1996, p 46)

It is interesting to see what Habermas has to say about the relation between the three models on the one hand, and citizenry and democracy on the other. According to the decisionistic model citizens elect politicians, but do not choose guidelines for ruling. Of course, a relation between particular politicians and a way of decision-making is assumed, but it remains to be seen whether this relation becomes actually true. "At best these decision-makers legitimate themselves before the public", Habermas says. "Decisions themselves, according to the decisionistic view, must remain basically beyond public discussion." (Habermas 1996, p 47) On the other hand, "[a] technocratic administration of industrial society would deprive any democratic decision-making process of its object." (Habermas 1996, p 47) For if politics should strictly obey to scientific standards, it is of no importance *who* is in politics. Here, it is assumed, falsely, that the degree of politics meeting scientific standards is not influenced by individual capacities of politicians, and that scientific standards are beyond dispute. However, here I will neglect these assumptions, and focus on the pragmatistic model. Habermas goes on to argue that

> the successful transposition of technical and strategic recommendations into practice is, according to the pragmatistic model, increasingly dependent on mediation by the public as a political institution. Communication between experts and the agencies of political decision determines the direction of technical progress on the basis of the tradition-bound self-understanding of practical needs. Inversely

it measures and criticizes this self-understanding in the light of the possibilities for gratification created by technology. Such communication must therefore necessarily be rooted in social interests and in the value-orientations of a given social world. (Habermas 1996, p 47)

What Habermas says here, is that technology develops corresponding to needs and wishes that exist against a particular cultural background. In fact, technology development and this cultural background interact, influencing one another by setting limits to what is feasible, and acceptable. Very importantly, however, Habermas maintains that "the value-orientations of a given social world" can only be made explicit through the public. Here, he implies that politicians and scientists cannot themselves account for them. If this is true, which I think it is, then RTA lacks a legitimate basis.

5.4
Rational Technology Assessment and its legitimacy

It is important to understand that with constructive Technology Assessment, evaluation seems to have lost its function of a judge. It is difficult to see how a feed-back mechanism, such as technology development linked up with insights into its accompanying socio-technological change, could be equated with assessment, if assessment aims at a judgement, or a strong indication: this way to go. Van Eijndhoven has argued that "the original, purely analytic character of Technology Assessment has been replaced by a view that gives credit to the process of Technology Assessment as much as to the analytical product." (Van Eijndhoven 1997, p 281) In the short history of Technology Assessment, an important shift can already be witnessed from Technology Assessment as an analytical activity towards Technology Assessment as a system of constant development, feedback, learning, moderation, and adaptation. Indeed, one could ask whether constructive Technology Assessment, including RTA, is still a kind of evaluation. Can it be distinguished from technology dynamics and technology development, or have these fields merged? Has Technology Assessment been disconnected from an Archimedean point that enables to take a critical perspective on technological change? Can we trust that a self-steering system of technology development, that is powered, rather than scrutinised, by Technology Assessment, accounts for its acceptability and justice?

To make matters still more complex, from a constructivist point of view Technology Assessment simultaneously scrutinises intervention efforts to improve our world, and represents itself such an effort. This is important. Not only is Technology Assessment applied in a social context, it is itself shaped by its social context, and reflects the norms and values adhered to in this context. Banta and Luce say: "It can be forgotten, in this rational and scientific age, that culture and society underlie all actions in health care." Of course, the same applies to other fields than health care. Banta and Luce continue:

"Technology Assessment [...] cannot be totally objective or value-free. As an activity carried out by human beings, it too is influenced by social and cultural values." (Banta a. Luce 1993, p 132) Choices as to which standards of performance are adopted, which criteria of merit are used, which stakeholders are included, which research questions are relevant, which outcomes are measured, etc., are determined by perspectives on what a technology-in-development is or should be. Problems may arise when stakeholders have different perspectives on a technology. Neglecting one or more perspectives results in a normative bias in the assessment (Reuzel *et al.* 1999).

What, then, can the legitimacy of Technology Assessment be derived from? Recent approaches to constructivism answer this question by seeking to let stakeholders actively participate in research and development, and policy making. (Apart from this, CTA seeks to use the unique knowledge of stakeholders.) Participatory designs and stakeholder empowerment are hot, so to speak. RTA, however, rejects these approaches on the basis of its specific understanding of rationality. But, of course, the very concept of rationality is a social construct, too. And what is more, it may very well be that it is the same rationality that has shaped our contemporary world as underlies the RTA design. Improving our world, then, is not merely a matter of introducing or emphasising "rationality" in Technology Assessment. Rather, it is this rationality that should be assessed before anything else.

The rationality that characterises RTA places RTA close to the technocratic model in Habermas' scheme, whereas CTA is more pragmatistic. With CTA, RTA has lost its original early warning function and its desire to judge technology development, and no longer primarily serves to improve public awareness. However, contrary to CTA, RTA deliberately has detached itself from the public sphere as an active mediator, too. Instead, RTA heavily relies on expert knowledge. No doubt, its interdisciplinary nature renders it an enhanced approach. Moreover, it is plausible to assume that a large expert group may set a fairly comprehensive agenda, and that no relevant issues are forgotten. Finally, the quality of work done by such an expert group, measured by this group's own standards, may be very high. Nevertheless, RTA lacks legitimacy. For it rules out the public sphere as an undesirable source of irrationality, whereas it is exactly this public sphere that should serve as a legitimate basis. The assumption behind RTA is that expert knowledge can establish itself outside the public sphere, that is, beyond need of legitimisation or justification. With Habermas, I would argue this simply *should not* be true. The road to stakeholder participation and empowerment that recent CTA approaches have taken, cumbersome though it may seem, may be a more justified one. We should meet this challenge, rather than leave it to the experts.

References

Banta HD, Luce BR (1993) Health care technology and its assessment: an international perspective. Oxford University Press, Oxford

Carpenter SR (1983) Technoaxiology: appropriate norms for Technology Assessment. In: Durbin PT, Rapp F (eds) Philosophy and technology. D. Reidel Publishing Company, Dordrecht

Gago JM (1998) The social science bridges. Portugese Ministry of Science and Technology, The social science bridge, Observatório das Ciências e das Tecnologias, Lisbon

Habermas J (1996) The scientization of politics and public opinion. In: Outhwaite W (ed) The Habermas Reader. Polity Press, Cambridge

MacKenzie D, Wajcman J (1985) The social shaping of technology – how the refrigerator got its hum. Open University Press, Milton Keynes

Rapp F (1983) The prospects for Technology Assessment. In: Durbin PT, Rapp F (eds) Philosophy and technology. D. Reidel Publishing Company, Dordrecht

Reuzel RPB, Van der Wilt GJ, Ten Have HAMJ, De Vries Robbé PF (1999) Reducing normative bias in health Technology Assessment: interactive evaluation and casuistry. Med. Health Care Philos. 2(3): 255–263

Rip A, Misa ThJ, Schot J (1995) Managing technology in society: the approach of constructive Technology Assessment. Pinter, London

Schot J (1992) Constructive Technology Assessment and technology dynamics: opportunities for the control of technology – the case of clean technologies. Science, technology and human values, 17: 36–56

Schot J, Rip A (1997) The past and future of constructive Technology Assessment. Technological forecasting & social change, 54(2,3): 251–268

Van Eindhoven JCM (1997) Technology assessment: product or process? Technological forecasting & social change, 54(2,3): 269–286

II Technology Assessment as Policy Consulting

6 Parliament, Paradox and Policy*

David Cope

6.1
Introduction

The paper presented here is based almost entirely on perspectives derived from the experience of the Parliamentary Office of Science and Technology (POST), an integral part of the UK Parliament at Westminster. As a consequence, it has certain characteristics:

1. Obviously, it primarily draws on the UK situation. Indeed a major aim in making the initial conference presentation was to exchange ideas that have emerged from this context that may be of interest to other experiences. Conversely, of course, we seek to learn from those other experiences any approaches that could potentially be used for the work of POST in the future.
2. It is set in the context of *Parliamentary* Technology Assessment (PTA). This shares a central body of method and process with all forms of Technology Assessment (TA) but, certainly in the UK situation, has some additional specific features. The most important are:
   - The results of any PTA must be comprehensible to a wide range of pre-existing familiarity with the subject of the TA within Parliament as well as enjoying the acclamation of any parliamentary 'experts[1]'. This acclamation is important for the following reason. Invariably among the 659 members of the UK Parliament's lower house – the House of Commons – and 300–400 'active[2]' members of the upper house – the House of Lords – there will be some 'experts' on the subject, with numerous perspectives. In all its TA work, POST aims to produce output which has the confidence of such people. They

---

* The views expressed in this paper are those of the author alone.
[1] Any parliamentarian who considers him or herself to be an 'expert' is so regarded.
[2] Although the situation has changed with the removal of the right of all but 92 heriditary members to sit in the Lords in 1999, numerous upper house members do not regularly take part in the activities of the House.

will then be more likely to trust its work in areas where they do *not* consider themselves 'experts' and also to commend its work to their colleagues. Beyond the reservoir of 'expertise' with Parliament, POST is a publicly funded body, so it is obliged to make all its work generally available[3]. It must therefore be ready for it to be scrutinised by the complete range of viewpoints on any issues associated with the subject, present in the UK population.

- The above circumstances are the foundation on which to claim that *parliamentary TA can be said to be inherently transdisciplinary,* and therefore of particular interest to those concerned with the subject of this conference.

- PTA work should not be prescriptive. This is very important in the UK context, where the PTA office is *within* Parliament and is employed directly by it. The drawing of policy conclusions is the prerogative of parliamentarians, not of the TA office. Direct recommendations are therefore not made. Instead, 'significant issues' are identified, using the term 'issue' in the sense in which it is used in English law, namely a matter about which there is contention. The outcome of pursuing different policy paths may be explored, however, as far as it is reasonable to do so, adopting an implict 'what if' approach.

- Self evidently, any TA output must focus on dimensions of the subject that 'resonate' with the role of the UK Parliament – the passing of legislation or the scrutiny of government policy. The output may however (and invariably does) include discussion of the limits of this role, highlighting where other actors, from the market to foreign governments or international agencies, will be significant influences. In as much as Parliament is invariably trying to assert its authority over government, the subjects chosen for TA examination and the emphases within a TA, tend to be oriented to assist Parliament in this perpetual 'arm-wrestling'.

- PTA must be both timely and succinct. The UK Parliament operates on a short term timetable between a subject appearing on the parliamentary agenda and the completion of its examination, at least initially. Two to three months is the normal period for a project, including external review of a publication draft. This constraint of timeliness is a major challenge and forces restriction of attention to the nub of the subject, as identified by the considerations of parliamentary interest. POST engages in a considerable amount of informed crystalball gazing to identify subjects likely to command the attention of parliamentarians in the future and to have its analyses completed, or well on the way to completion, before their parliamentary consideration. Succinctness is as great, if not a greater, challenge to report

---

[3] The external dissemination of POST's work for parliamentary committees is at their discretion, although POST invariably seeks that this be done.

authors – parliamentarians' time for in-depth reading is very limited.
The majority of the office's published output consists of four page
briefings, though longer reports, conforming to the 'orthodox' format
of TA, are also produced. Brevity is a very strict disciplinarian, again
forcing attention solely on the nub.

- In general, the time-frame over which subjects chosen for PTA are
examined is short. Parliamentarians have limited interest in strategies
which extend over the lifetime of more than one Parliament (five years
at most), while, more generally, there is currently within the UK polity
an intellectual predisposition to eschew grand long-term planning.

Petermann (2000), places POST and its work as lying on the 'instrumental' side of an instrumental/discursive dichotomy that he uses to describe the
varying approaches of European PTA organisations. No-one could gainsay
that but, as he notes, the dichotomy is not absolute. POST has been 'developing a relationship' with various policy consulting procedures in recent
years. Furthermore, it has recently been requested, by a parliamentary committee, to take on the specific role of 'keeping Parliament informed' about
wider use of consultative techniques in science and technology policy.

Before turning to discuss POST's various experiences in TA consulting,
I would like to set the subject in the context of two wider 'paradoxes' and
explore whether parliamentary TA policy consulting has role in resolving
either of these.

6.2
First paradox: 'The Paradox of Parliamentary Democracy'

A widely held view in political science is that liberal democracy (i. e. implicitly involving a parliamentary system) has triumphed globally. The most
refined exposition of this thesis argues that this form of government is the
'end point of mankind's ideological evolution' (Fukuyama 1992). The paradox
is that, at the same time that this process has been playing out, participation
rates in conventional democratic processes, such as the proportion of citizens
voting in parliamentary and other elections, has been declining. This is certainly true in the UK, especially for parliamentary by-elections. Beyond this,
there is argument in the UK that the statuses of Parliament itself, of the political process and of the politicians who participate in it are also diminishing
(Riddell 1998 and 2000). At least two unofficial 'commissions' have been set
up to examine this paradox and to recommend ways of stemming the decline.

In the UK, there are probably some specific reasons that have accentuated
this paradox, arising from political events in the recent past:

- The image of politicians has been tarnished by scandals over financial im-
propriety and a belief that at least some of them have exploited their
political position for personal gain.

- The government has had a very large majority since 1997; therefore to some extent politics has less intensity than in the period before this.
- Some argue that it has been the 'philosophy' of successive governments to subdue Parliament – to constrain its role in scrutinising government activity. With the current government, they point to its apparent unwillingness to bolster the powers of parliamentary committees (including extending their research capability). They further accuse it of 'foot-dragging' over reform of the second chamber – the House of Lords – particularly regarding the possibility of at least a proportion of members being directly elected.

Is there a resolution of this paradox? The US political scientist, Dahl (2000), maintains that democracy has an inbuilt duality. It consists of a body of rights and opportunities (institutions) and, secondly, the actual exercise of those rights – participation. Dahl believes that the 'first dimension' 'may well be' the more important. He argues that citizens attach a very high 'option value' to the institutions of democracy but may not actually seek to participate, wishing rather to do other things in their lives. He sees this as a perfectly acceptable position.

Dahl however, does note that dissatisfaction with the way their government works might in the long run reduce the confidence of some citizens in the value of the first dimension of democracy and thus weaken their support for democracy overall. However, as compensation, he believes that others may be driven to participate more fully, in a bid to rescue democratic ideals. He does not explore the fine dynamic between these two competing processes, while there are those who argue that it is too dangerous to rely on the spontaneous emergence of a self-correcting reaction.

6.3
Second paradox: 'Science, Technology and the Public'

This has some parallels with the first paradox – for democracy, substitute 'science and technology'. It is encapsulated in a frequently asked question, namely why is it that by some measures (including government spending and interest in popular science, as indicated by sales of books, the popularity of TV programmes, etc) science and technology are enjoying a golden age – but at the same time they are said to be losing public confidence. As evidence of this loss, people point to apprehensions over aspects of biotechnology or the future of civil nuclear power.

I deliberately use the phrase 'said to be' in the paragraph above because public opinion research evidence is inconclusive. It is, however, undoubtedly true that a large body of media coverage argues this is the case, criticising, among other things:

- The independence of scientific research – that much research is now in the private sector and driven by that sector's overriding concerns. Because many public research institutes are increasingly relying on private sector contracts, they are similarly seen as incapable of providing neutral advice. It remains to be demonstrated that advice from such sources is *invariably* partial, while pressure groups, sometimes held as untainted in this respect, have been known to present selective or distorted information to further their own ends.

- the 'peer review' system for verifying publication (and deciding the direction of science and technology research through grant procedures). This is said to be flawed through being too exclusive – dominated by a self-perpetuating set of established individuals unwilling to consider wider perspectives or to support research which does not conform to their own paradigms.
- the hypothetico-deductive 'scientific method' itself in some cases. Some quarters argue that it is but one of a set of 'models of existence' or 'values', others of which have equal or greater merit (needless to say, the value set which those who stridently argue this case advance as having the greatest merit is invariably their own). Allied to this, (a 'weaker' interpretation in the technical sense) is the argument that there are boundaries to the rule of science and some areas where it has no sway (or at least should not dominate).

Despite this, as noted, many governments throughout the world, and certainly the UK government, are placing a very strong emphasis on the wealth creating role of science and technology, including moderate amounts of increased public expenditure. Although clearly welcome to the science and technology community, this trend might have some downsides:

- obviously, science and technology must deliver the 'goodies', (and quickly) lest disillusionment sets in. There is a risk that too much may come to be expected of science and technology.
- the lionisation of science and technology and its identification as the motor of wealth creation has attracted the attention of schools of thought which, to a greater or lesser extent, oppose the current economic system and/or wealth creation. The rationale is that if science and technology underpins these, then attacking science and technology will weaken them. Undoubtedly such a *weltanschauung* exists (held by a protean and amorphous alliance in the UK). Unfortunately, the recent successes of science and technology in areas such as genetic modification and human genomics, which impinge on matters where the general public almost inevitably have atavistic concerns, has given considerable influence to such schools of thought.

6.4
Policy consulting in contemporary context

Two questions spring to mind. The first is simply – why consult? This question might seem redundant, if not downright impertinent. Is not consultation an integral part of parliamentary democracy? If it is, then the second question arises, namely who is to be consulted, when, and about what?

To answer, I think it is helpful to identify two types of consultation – I call them *instrumental* and *deontological*:

- *Instrumental consultation* is conducted to get better results for an assessment exercise, to make certain that as many as possible of the knowledge sets, all the perspectives, etc, that exist, are known to the assessment researchers and can be incorporated in the results of the assessment. An associated rationale of this instrumentality is to alert the consultees that the work is occurring, to promote their cooperation in supporting dissemination of the output, to build up awareness and the reputation of the assessment institution, and so on.
- *Deontological consultation* is conducted because consultees have a *right* to be consulted, independent of the utility of the results of the consultation for the assessment exercise. A right to be consulted exists in some areas of UK law (e. g. town and country planning) but not routinely in the TA field. The UK government invariably consults (by issuing consultation papers, inviting responses) about new policy proposals, although it is under no legal obligation to do so. Recently, however, it has written into the terms of reference of several new advisory/regulatory bodies[4], an obligation to consult – these bodies are now exploring how they should discharge this obligation.

The current UK government made a great play about consultation when elected in 1997 (although again it was under no legal obligation to pursue this path). For example, it set up a 'Peoples' Panel' – a very large 'focus group' of citizens, with which to explore major policy areas. More recently, there has been some evidence of disillusionment by government about the utility of this initiative (and considerable attrition of Panel membership). Beyond this, there has been some more general discussion in the media of 'consultation fatigue' – a feeling that there has been too great a resort to consultation, some of it ritualistic rather than signifying any real change in policy-making. The debate also draws on discussion of the relative merits of political 'leadership' versus the pursuit of consensus, to which the adjective 'flabby' is often attached, explicitly or implicitly.

---

[4] Such as the Human Genetics Commission, the Food Standards Agency and the Agriculture and Environment Biotechnology Commission. For example, the terms of reference of the last require it to 'seek to involve and consult stakeholders and the public on a regular basis on the issues which it is considering'.

6.5
Some policy consultation experiences of POST

6.5.1
'Electronic'

The growth of the internet has undoubtedly made it easier and cheaper to conduct various forms of consultation, obviously limited in their outreach only to those who have internet access. These tend to be 'experts', although as internet access becomes more ubiquitous, this restriction is diminishing.

POST has used pre-existing e-mail subject groups (without any in-depth prior investigation of the membership of the group) to consult on project outlines and to provide specialist expertise. Obviously this is expert consultation, although the range of group membership, from graduate research students to senior academics and administrators, can be attractively wide.

In two cases, POST has itself created temporary ad-hoc e-mail subject groups. The first was an expert consultation to discuss a then current piece of legislation (on data protection). Here, POST 'snowballed' the membership (i.e. it invited a few initially-invited participants themselves to nominate other participants). The second consultation involved using widespread publicity through the internet but also in the press and broadcast media to invite participation in a specially-created e-mail group to discuss the educational and career experiences of women in the fields of science and technology (membership was not limited to women). Here, the intention was to consult with a group having a particular *experience* – no particular 'expertise' was relevant. In fact, an explicit intention was to go beyond the 'noisemakers' in this area to get at the experience of 'ordinary people'. The results of this consultation were fed into an ongoing parliamentary enquiry on 'Science and Society' (House of Lords 2000).

6.5.2
'Non-electronic'

POST has recently tapped into a pre-existing expert network, the Pugwash Conference, to ask participants in a Pugwash workshop to referee[5] a Technology Assessment on ballistic missile defence systems. The particular attraction of this approach lay in the quick and easy access it gave to a wide range of international perspectives on a subject with profound international dimensions.

Perhaps, however, of greatest relevance to the interests of this conference is that POST has had involvement with two national (the only two, to date) 'consensus conferences' in the UK – the first, in 1994, on plant genetic modification and the second, in 1999, on radioactive waste management. As is

---

[5] All POST publications are externally refereed in draft form.

well-known, the explicit aim of consensus conferences is to go beyond expert consultation to reach out to 'laypeople'.

It should be noted that the text above says that POST has had 'involvement with' the consensus conferences[6]. Unlike some of our sister institutions elsewhere in Europe, POST did not itself organise the consensus conferences and has no intention of doing so in the future. Cost constraints are a major consideration[7], but above this there is a political philosophical principle – that it is not the responsibility of an integral part of Parliament so to do.

Again, unlike some of our sister institutions elsewhere in Europe, POST has no remit to articulate public debate, to inform the wider world beyond Parliament[8] or to seek to determine the 'public mood' on any particular issue. Regarding the last, it is, of course, traditionally Members of Parliament themselves who are the channel for this, through the soundings they make in their constituencies and through the contents of their mailbags. Some parliamentarians could become upset if they felt that this role was being compromised by a body from within Parliament itself. Indeed there is a body of opinion, within and outside Parliament, which is unsympathetic to public consultation procedures no matter by whom they are conducted. This body draws on concerns about 'government by referendum', populism, etc[9]. Equally, however, there are those who are enthusiastic about the extension of consultation. Reflecting the recent debate on such approaches, the House of Commons Public Administration Select Committee has recently held an inquiry, expected to report before the end of 2000, on *Public Participation: Issues and Innovations*.[10] The key indicator will, however, be the content of the government's response to the committee's report.

The overall positioning of POST in the context of public consultation that is outlined above received official parliamentary endorsement earlier this year in the report of the House of Lords Science and Technology Committee on *Science and Society* (House of Lords 2000). This commended us for not embracing uncritically what it interpreted as a 'Danish model' but, on the other hand, gave us a remit to keep Parliament informed about experiences in public consultation on science and technology policy.

---

[6] In both cases, POST was a member of the consensus conference steering committee and with the second, it also hosted the final deliberations of the lay panel prior to public presentation of their findings, at its offices.

[7] The budget for the 1999 consensus conference was over £ 100,000, the lion's share of this being the costs of bringing lay panel representatives from across Britain to two preliminary weekend meetings and the final three-day deliberation and findings presentation in London.

[8] All POST publications for parliamentarians are, however, also made publicly available in printed form and on the internet.

[9] For a discussion of the broader political philosophical issues involved, see Manin (1997).

[10] Finally published on 5 April 2001 (HC 373 I and II)

6.6
Criticisms of consultative techniques

From its experience of consultative processes, POST has been confronted with several criticisms of procedures, especially of exercises designed to reach out to the 'general public'. Among the more significant are:

- *Representativeness* – The process for selecting consultees may not be random. Clearly this is not an issue when the aim is to select experts for expert consultations but it certainly applies to consensus conferences and similar tools aimed at revealing the opinions of the citizenry. The most effective way of approximating a fully representative cross-section of the UK population is to take a sample from the electoral registration lists, which in theory contain all those eligible to vote (i. e. UK and Irish citizens and EU citizens for European elections). In fact, there is under-representation, particularly in some parts of the country. Even if the selection frame is representative, criticism is made that those who actually agree to participate almost inevitably are not representative. For example, they may need to be willing to give up weekends, which may militate against citizens with young families. It is difficult to disagree with this observation. Beyond this, some say that a further bias is that those who respond positively are more likely to be 'participators' anyway, i. e. to have a psychological predisposition to get involved. Fishkin (1991 and 1995) argues that, even if true, this is not an issue, because one aim of consultative procedures is to promote deliberation on policy issues. From this promotion emerges an outcome which is that which would arise if deliberation had occurred more generally among the population as a whole. The inference is that the deliberative outcome is 'better' than the snap decisions of the general population. It is difficult to deny, however, that politicians may be more influenced by the dominance of such snap decisions, however morally worthy might be the goal of promoting deliberation. There is also the question of whether the deliberation of people who probably are predisposed to be 'participators' will indeed produce the same results as the hypothetical deliberations of those who are not so predisposed.
- *Replicability* (robustness of outcome) – The argument here, again applying particularly to citizen consultations but also to expert consultations where there is no guarantee that *all* experts have been drawn into the process, is essentially a statistical one – that because the consultation relies on only one sample from a much larger population, no robust conclusions can be drawn from the outcome. The best way to test this would indeed be to conduct a multiplicity of consultations and then to compare outcomes. This would, however, be virtually impossible with consensus conferences, to which this criticism has been particularly addressed, because they are, as noted, so expensive to run.

- *Citizen competence* – This matter of contention, again applying solely to citizen consultation on science and technology issues, is part of a wider debate on the boundaries of 'expertise'. The accusation is that citizens, lacking specialist knowledge bases, are completely ignorant about complex technological issues, so that these are inappropriate for handling by consultation techniques. For example, in the field of radioactive waste management, there is scientific uncertainty about, and therefore dispute over, flow rates of water in deep rock interstices. Critics of plans for the 1999 consensus conference ridiculed the idea that ordinary citizens might resolve this uncertainty. This argument, however, derives from a fundamental misunderstanding of the aims of consultation, which are not for the public to 'adjudicate' on matters of scientific uncertainty but rather to establish exactly what are the areas of public concern and to illuminate the policy actions that might meet these concerns and resolve outstanding policy issues.
- *Ephemerality* – The argument here runs that even if the results, especially of citizen consultations, but maybe even of some expert group exercises, are indeed a measure of the 'public will' or of expert consensus, they will probably have such a limited 'shelf life' as to make teasing them out rather nugatory. In particular, public opinion can change very rapidly in response to individual events, so that trying to establish what it actually 'is', is like chasing a will-o-the-wisp. Again, there is some truth in this assertion but not all consultation outcomes will rapidly date in such a way.

6.7
Conclusion

I will conclude by examining briefly some dilemmas in defining what is a 'successful outcome' in policy consultation, especially consultation which has deontological elements, as defined in the dichotomy outlined in section 6.4 above.

Sometimes, science and technology interests, while claiming to embrace the idea of public consultation in scientific and technological fields, implicitly or explicitly reveal that 'success', as they understand it, will occur only if, as a result of the consultation, at least an acquiescent population will emerge, enabling them to continue business as usual. 'Overwhelming success' would be not just an acquiescent population but an acclamatory one which lionised the achievements of science and those who practise it. Conversely, for others 'success' would be the adoption of a voluntary moratorium on the pursuit of some field of scientific enquiry, with 'overwhelming success' being a governmental, or better still, inter-governmental, ban on the same pursuit. In between these two extremes lies a continuum of modifications to current practice and institutions.

Politically, however, success can be defined rather differently as the re-moval of the issue from near to the top of the political agenda, even though it may never go away entirely, so that governments are not under pressure to act on the issue and can get on with doing other things that they want to pursue. Politically, therefore, there is less concern about reaching the 'correct' resolution and more concern with the expeditious achievement of the *easiest* resolution. I would argue that it is one responsibility of PTA to try to bring these two different goals into congruency.

Currently, the main perceived weakness of policy consulting in the UK is a lack of what I call 'follow-through'. Those with legislative or adminis-trative responsibility for an area explored in a policy consultation have been extremely reluctant to commit themselves, in advance, directly to pursue the policy outcome that emerges from the consultation. They make flatteringly placatory noises[11] – that the consultation is 'very interesting', that 'they will certainly take note' and so on but that is as far as they are prepared to go. They are not, in fact, willing to allow a transfer of power from themselves to 'the public interest'. This fear is perhaps understandable, for if they were to do so once, then they would almost certainly be under pressure to do so frequently, if not always. Public opinion on radioactive waste management may be broadly attractive to the UK government, but that on fuel taxation certainly is not!

Interestingly, the same reluctance has also been true of campaigning groups, who tend to refuse to modify their position if it does not mesh with that emerging from the consultation. That is, of course, their prerogative, because they are attempting to change 'public opinion' but it is a weakness which some of the more perceptive of pressure group leaders have recognised, causing them to be noticably hesitant about supporting public policy con-sultation exercises. Others attempt to emasculate them by raising spurious objections about their impartiality.

In the UK, I suspect that the future of citizen policy consultation initia-tives probably depends more on the experience of its use in policy areas other than science and technology (e. g. pensions provision, or the broad principles of taxation policy). I have already referred to evidence of an emerging 'con-sultation weariness'. Perhaps this is a sign of an inchoate 'return of trust' to experts. Undoubtedly, some big tests of this entire scenario will arise in the near future, for example, if serious consideration emerged of building new nuclear electricity generating plant in the UK, as is beginning to be talked about in some circles, including those outside the industry. That, however, merits a separate paper.

---

[11] Indeed, who could be *against* the wider consultation and involvement of the public – surely the quintessential 'motherhood and apple pie' of policies?

References

Dahl R (2000) A Democratic Paradox? Politicial Science Quarterly 115, 1

Fishkin J (1991) Democracy and Deliberation. Yale University Press, New Haven

Fishkin J (1995) The Voice of the People, public opinion and democracy. Yale University Press, New Haven

Fukuyama F (1992) The End of History and the Last Man. Hamish Hamilton, London

House of Lords (2000) Science and Technology Committee, 3rd Report, Session 1999–2000, Science and Society. HL 38 Stationery Office London

Manin B (1997) The Principles of Representative Government, Cambridge University Press

Petermann T (2000) Technology Assessment Units in the European parliamentary systems In: Vig NJ, Paschen H (eds) Parliaments and Technology, the Development of Technology Assessment in Europe. State University of New York, Albany, pp 37–61

Riddell P (1998) Parliament under Pressure. Gollancz, London

Riddell P (2000) Parliament under Blair. Politico's, London

7   The Politics of Technology Assessment.

Miltos Liakopoulos

7.1
Introduction

Technology Assessment (TA) is not a political issue, or to put it more accurately, should not be a political issue. The word 'political' here is meant as an ideologically charged debate whereby socio-economic theories or viewpoints dictate the general direction of the solution of the debated issue. TA should be protected from the ideological bias that is evident in political debates. Even if there is an ever increasing debate over the 'politicisation' of science, this should pose an independent issue to that of TA. This forms the general thesis of this paper.

The previous author gave an account of how TA functions within the UK national parliament. Despite its particularities, the UK parliament is not very different to other European parliaments that also have their own offices of TA. In that, there are some common experiences in parliamentary offices of TA that inevitably revolve around political issues. The issues raised by the previous paper are inexorably connected to the work of political institutions as they influence Technology Assessment. The comment that follows is based directly on these issues while trying to present them under a different light. There are three main issues that are commented here: the limitations that parliamentary offices of Technology Assessment are faced with, the methodological process of interdisciplinary exchange and consensual decision making, and lastly, the phenomenon of decreasing public trust in political institutions.

7.2
Parliamentary offices of TA: the limits

The first issue raised by the previous author was a general problem parliamentary offices of Technology Assessment are faced with: as parliaments are functioning as the main debate forum for various and often contradictory socio-political views, any institute affiliated with them and pertaining ideology-free work, is bound to be faced with considerable constrains. The experience of the British parliamentary office of TA is indicative of this situation.

We are told that there is a need to make considerable compromises between various forces; the work of a parliamentary office of TA seeks both scientific as well as political acclamation for its work. As some members of the parliament can be highly specialised in various scientific disciplines, holding their own scientific views, there is already "in-house" expertise that the TA study should satisfy. But more controversially, any TA study by parliamentary offices has to acknowledge the main political perspectives in the issue under consideration. The word "acknowledging" here means including in the study mainstream political views in a balanced and non-partisan manner to avoid accusations of bias or taking sides. This represents a precarious balance which creates one of the main obstacles in the work of parliamentary offices of TA.

In the world of politics ideological differences are the "reason for being" of political parties. These differences are the result of different interpretations of the same facts. The ensuing debates are won over by arguments which create dominant ideologies in society. Many people suggest that science is not different in the sense that it is also based on subjective interpretations of facts and is influenced by the dominant ideology (e. g. Lewontin 1991). Moreover, decisions on research funding are highly political and create a "one way road" for scientific discoveries where alternatives are out of question simply because they offer no political advantage[1].

This debate is important and should of course be pursued further. If there is a need for another scientific paradigm that better fits the new era of political awareness (Funtowitz a. Ravetz 1993), we should all gain by that. Nevertheless, when it comes to assessing new technologies, there should be no debate over "political after-maths" of scientific discoveries. In other words, parliamentary offices of TA should be able to assess new technologies without considering political views or political side effects. That is the only way parliamentary TA can keep the necessary integrity to provide valuable policy advice. The work of the parliament will inevitably discuss all political issues involved in the scientific advances under consideration.

Of course that is easier said than done. In reality parliaments are the designers and paymasters of such offices and as such, they can have considerable influence over their functions. The temptation to patronise the outcome

---

[1] One could here mention the example of financing research on genetically modified crops versus poly-croping (alternating crop cultivation on the same land). Some people claim that although both alternatives can prove equally successful in raising productivity, the latter does not receive any research grants as it confers no political gains (New Internationalist).

or simply use the office as a "scapegoat" might prove too big to ignore[2]. How can one then guarantee the independence of the work of parliamentary offices of TA without compromising their value as policy advisors? One way is to create a wide interest governing board that would include both parliamentarians and representatives of a variety of other interest parties[3]. That nevertheless, might prove very difficult for big and heterogeneous societies where ideological differences permeate most social segments (that is, most European societies).

Another scheme could view complete independence for parliamentary offices of TA in the manner of other independent TA institutes[4]. That would increase immensely the credibility of their work and would also give them the opportunity to work on in-depth studies with a view to long-term policy advice. For example, the previous author reminded us that the majority of their studies is not more than four page long briefings. Although everyone would naturally tend to agree about the need for succinct scientific studies that are readable by busy professionals (that is, these days, everybody), studies of that length might border the impossible. Short briefings produced within few weeks cannot be expected to provide meaningful policy advice or input in public debates.

7.3
Interdisciplinarity and Consensus: possible, desirable or simply necessary?

Another issue referred to by the previous author is that of the "transdisciplinarity" in TA studies. "Transdisciplinarity" in that context refers to the effort to include the main scientific perspectives on a particular issue, especially in view of the parliamentary in-house expertise. This of course represents a positive step toward a more integrative and inclusive analysis which is the aim of TA but it is not perhaps the best process available.

A more useful method is to be found in the model of interdisciplinary projects of TA[5]. In this model, the report is compiled by a group of experts representing different disciplines relevant to the technology in question. There is no absolute rule as to the number of disciplines that can be included in such project but they can represent both "core" as well as "peripheral" aspects. For example, a health care technology project might involve experts from medicine, physiology and engineering, but also ethics, health psychology,

---

[2] One can give as examples the closure of OTA in the USA or the turmoil that the ECHELON study created in the European Parliament.

[3] One such example is the Danish Board of Technology whose governing board includes members of trade unions, research councils and local authorities.

[4] The Europaeische Akademie is an example of a publicly funded institute with complete independence over its manner of work and choice of themes.

[5] For a detailed description of such model, see Decker and Grunwald (chapter 4).

health economics and even public perceptions of new technologies. Such wide range of interests adds naturally more complications in the process but can also result in a more "pragmatic" policy advice. The term "pragmatic" advice in this context refers to the fact that recommendations from a particular disciplinary perspective will be automatically reviewed and commented upon from a variety of different perspective in terms of its viability. It is therefore important to create a continuous process of interchange between disciplines with the ultimate aim of a synthetic view.

Indeed, the interdisciplinary model of TA can bring considerable obstacles during the process of the exercise. Experts in particular disciplines have naturally only rudimentary knowledge in other disciplines, and the further apart the disciplines the more peripheral that knowledge is; one could then argue that they are no more than "educated lay persons". In this situation, it takes substantial effort to follow and understand state-of-the-art knowledge in other disciplines that might be related to their subject[6]. But the result is no less than a better understanding of one's own discipline and its wider consequences.

Naturally, the ultimate purpose of this great effort in widening one's understanding in a TA study, is to be able to provide policy makers (who still hold the ultimate responsibility for decisions) with meaningful advice which is based on some form of consensus. The simplicity of the word hides a universe of complexity and inspires heated arguments. What is consensus and why should one aim for it? Nowadays "consensual decision making" poses as a policy ideal that is tried in virtually every international agreement of significance. From the UN system to the EU, every major organisation requires consensual decisions by all its members (which usually represent very different perspectives). It is of course impossible to provide even a sketch of the experience of consensual decision making here. Our question, in relation to TA, is not how easily achievable it is but rather whether consensus is desirable or not. The answer, I believe, is a simple "yes".

In interdisciplinary TA continuous interchange and reviewing amongst the various disciplines and perspectives, creates a formalised process of Technology Assessment that has been termed 'rational Technology Assessment' (see Grunwald 2000). This form of TA is claimed to suit better the needs of long-term policy planning while differentiating between "acceptance" and "acceptability" problems (the latter of the two being its main focus). While there is an on-going debate on the various methods of TA – this conference being another part of it – it is wise to provide only a tentative suggestion: rational interdisciplinary Technology Assessment leads to consensual decision

---

[6] For example, it might require substantial cognitive effort for a food biotechnologist to understand the effect of his discipline to the balance of payment in developing countries as he/she would have to understand the principles of scale-economies.

making. Rationality assumes approval of common set of communication rules that, given enough time for appropriate exchange, should lead to agreement.

This might be counter-intuitive to many people who view many uncertainties in scientific research. Truly enough, the scientific paradigm of hypothesis testing allows for a wide view of the "cause and effect" relationship. Statistical significance is the current test of truth for the hypothesis but at the same time carries an in-build error that is often filled with alternative hypotheses that might be opposite to the original one. Further complications arise when the hypothesis enters the realm of socio-economic reality which usually provides further hypotheses.

Nevertheless, despite the difficulties inherent in the scientific process, the need for decision making is real and unavoidable. In an interdisciplinary process we find the ideal situation to reach an agreement over pragmatic recommendations. The unavoidable gaps in scientific knowledge can be filled by the appropriate input by other disciplines thus, creating the basis for consensus. It should be clear that any consensus in scientific matters is conditional upon future developments that might shift the direction of the recommendation. That does not only diminish its initial value in policy advice but it even increases it as a platform for procedural exchange. An interdisciplinary consensus creates a formula that can be revisited and used as the current policy needs dictate.

7.4
Trust: ubiquitous by its absence

The previous author also referred to another important aspect which represents a recent phenomenon: the decline of public participation in democratic processes, followed by the decline of public trust in government bodies and, more importantly for our inquiry, the scientific establishment.

There are indications that there is diminishing interest in electoral processes (e. g. record lows in election participation rate) accompanied by declining trust in the work of parliamentary bodies. This trend has been evident all over Europe especially in relation to socially sensitive scientific issues. For example, in a recent European survey on attitudes to biotechnology, where respondents were asked which institution they trust most to tell them the truth about that new technology (the list covered a range of NGOs, public and private institutions), only 2.7 per cent of the European public choose their national government bodies as the most trustworthy institutions (Eurobarometer 2000).

There are also indications, albeit less clear, that there is a similar situation in the area of science and technology (S&T). On one hand there is an overall evident trend towards declining trust in S&T information (Jespen 2000) and on the other hand there are specific issues that witness the opposite effect[7]. Despite the uncertainty over the whole picture, the issue of trust in S&T has become a major political theme with calls for the re-evaluation of scientific political decision making and the need to follow more socially accepted models[8]. This major shift in science policy is partly due to the failure of the so-called "deficit model" whereby the interaction between science and the public is seen as one-way process of "public education" (see Dierkes and von Grote 1999).

The debate on public trust, its dimensions and direction, is of course important for the future of the political system. But despite the severity of the situation for policy makers, this debate should have little if anything to do with TA. These worrying trends we are witnessing in participation are purely political issues that should not affect the manner in which assessment of new technologies is done. The previous author, for example, attempted an explanation of the declining participation in parliamentary democracy in the UK by referring to the particularity of the UK political situation; in this way, he rightly set local limits to an issue that consequently merits local solutions. Similarly, the 'paradox of parliamentary democracy' in other countries merits different analyses and possibly different solutions.

Nevertheless, there seems to be an interesting association between certain TA procedures and the general lack of public trust. There have been claims that public consultation in TA issues is not only a way to understand better the consequences of new technologies, but it is also an experiment in direct democracy that could perhaps help to recover the lost public trust (e. g. see Durant 1995). This argument has helped making consultative techniques the mainstream TA approach. Despite their popularity, these techniques seem to be coming to a crossroad signified by "consultation fatigue" and government disillusionment with the practice (as the previous author pointed out). Despite that, the British government might have paved the future way in how modern societies deal with science problems, by including the obligation to consult the public in the terms of reference of many newly created regulatory bodies.

Taking that as a point of departure, I think there is little doubt that in a democracy, public consultation is important; politicians need to know what the public think about important issues and the public also need to feel that they are being heard. But, could public consultation be used as a political instrument and with a political aim? The previous author slightly referred to claims that consultation does not really lead to policy change. One could also say that there is no conclusive evidence showing consultation techniques

---

[7] For example, the same European survey on public perceptions to biotechnology showed a considerable increase in public trust towards scientific organisations compared to the previous survey in 1996 (Eurobarometer 2000)).

[8] Indeed, there has been recently a major EC funded conference on 'science and governance' that signalled major changes in the future culminated by the proposal to create a new European Commission Directorate General dedicated to the issue.

having any impact on policy making. Sceptics might argue that governing bodies have every reason to appear to be consulting the public without any commitment to take on the advice.

Given the apparent obscurity surrounding the use of consultative techniques in TA, one is compelled to question its overall use in TA. Should public consultation be part of TA? Or, to put it in other words, is public consultation necessary for TA? The answer to both questions could be conditional to the situation. In many cases S&T issues are of a general nature and refer to widespread effects, while in some other cases they refer to local situations that require local solutions. It is more reasonable to use consultative techniques in the latter case but not in the former. The reason is that in the latter case consultation becomes by definition a necessary and significant policy input. In technical problems affecting specific localities, local residents can provide valuable information as they will be the most directly affected by the ensuing changes.

In the case of scientific and technological issues of widespread consequences, consultative techniques can be of very limited value. There are some well known problems of representation: who should be asked to be part of the consultation process from a potential pool of millions? Then, consultation techniques require considerable political commitment that seems to be lacking nowadays; Can the results be used in policy making in any case or unconditionally? Lastly, when it comes to analysing technological advantages and disadvantages with a long-term perspective, one needs unavoidably an increase in cognitive capacity[9] that requires exchange processes identical to those of expert or rational TA. It is no surprise then that results of consultative exercises on wider issues have been praised as extremely rational but offer little value to policy makers[10].

7.5
In Conclusion

This paper gave a brief comment to various issues raised by the previous author that are by any means pivotal in the debate on Technology Assessment. The limitations of parliamentary TA have been claimed to be the result of the closeness between the two institutes (successful relationships are never based

---

[9] For a more detailed conceptual analysis of the role of cognitive capacity in TA see C. F. Gethmann, chapter 1.

[10] For example, the first UK consensus conference on "plant biotechnology" in 1994 resulted in a very balanced view of the subject that was highly praised by the scientific authorities present. The qualifying support to plant biotechnology by the lay panel (the majority of which said they were satisfied that genetically modified foods pose no threat), gave policy makers a completely wrong picture of public views of the technology. It turned out that the UK public is one of the most critical in the world in issues of GM crops and foods.

on dependencies). Moreover, the issue of interdisciplinary research in TA and consensual decision making have been argued to be inexorably connected; interdisciplinary processes assume a pragmatic rationality that in turn leads to consensus. Lastly, the very current and "hot" issue of public trust has been related to that of consultative techniques in TA with the claim that declining public trust is mainly a political issue that requires political solutions; there is little TA can, or should, do for that.

Concluding, one should remember that Technology Assessment is young (and parliamentary Technology Assessment is even younger). The debate over its methods and practices is still rife and healthy. The variety of techniques used or tried out is indicative of the remarkable creativity surrounding the subject. There is doubtless more knowledge to be gained and experiences to be made. For example, despite the different methods being used for considerable time (some of them for over ten years now), there is very little research on the actual impact of TA techniques in decision making, and there is certainly no team action to identify and standardise impact assessment procedures. This could be the next important step in TA methodological research that would also probably provide many answers to questions over the value of its different techniques. Perhaps then, the debate on TA can reach another, more conclusive level.

References

Bauer M (1998) Biotechnology in the Public Sphere: a comparative review. In Durant J, Bauer M, Gaskell G (eds) Biotechnology in the Public Sphere. Science Museum, London

Dierkes M, von Grote C (1999) (eds) Between Understanding and trust: The public, science and technology. Harwood Academic Publishers, Reading

Durant J (1995) An Experiment in Democracy. In Joss S, Durant J (eds) Public Participation in Science; The role of consensus conferences in Europe. Science Museum, London

Eurobarometer 52.1 (2000) Europeans and Biotechnology. INRA, Brussels

Funtowitz S, Ravetz J (1993) Science for the Post – Normal age, Futures, 25, 7, 735–755

Grunwald A (2000) Technology Policy between Long – Term Planning Requirements and Short – Ranged Acceptance Problems. New Challenges for Technology Assessment. In: Grin J, Grunwald A (eds) Vision Assessment: Shaping Technology in 21st Century Society. Springer, Berlin

Jensen P (2000) Public Trust in Scientific Information. IPTS, Seville

Kluever L (1995) Consensus Conferences at the Danish Board of Technology. In: Joss S, Durant J (eds) Public Participation in Science; The role of consensus conferences in Europe. Science Museum, London

Lewontin RC (1991) The doctrine of DNA. Penguin Books, London

III Participatory Technology Assessment

8 Toward "lay" participation and co-operative learning in TA, technology policy, and construction of technologies

Imre Hronszky

Abstract

The paper first outlines some features of the historical development of TA, relevant to the problem of citizen, "lay" participation. Following this it makes some remarks on the changing role of technology in society. Then I make some remarks about models of political democracy and consequent preferred ways of organising cognition. This section also introduces the claim for "technological citizenship" as a normative political consideration for appropriate participation in processes of the political regulation of technological development. The idea of "technological citizenship" will be discussed with relation to raising questions about the requirements necessary for experts and participating citizens as "lay" persons making informed contributions in a participative Technology Assessment. The main part of the paper is a critical assessment of the cognitive potentials and limits of the "lay" public. As a conclusion the need and possibility of developing a mutual learning process will be emphasized.

An introductory note on the "moral" intended by the paper is also in order. Some recent actors in the TA arena still continue to play down the role of either the experts or the "lay" public as appropriate, essential cognitive actors in making TA. In doing so they support a move in a "downward spiral", that is in a trajectory along which mutual "denigration" of the expert and the "lay" ruins any possibility of rational co-operation. The paper tries to show that both experts and the "lay" public are necessary actors if an "upward spiral", a reinforcing co-operation and with this a higher level of rationality in TA issues is intended to be realised.

It is more and more acknowledged and the opinion reinforces that "lay" persons, the citizens, have a role in assessing technologies. However, arguing for the essential participation of "lay" persons already divides the parties. Recently the debate has moved to being more about whether they should be involved as political actors or also as, in some respect, essential actors in the cognitive process. My understanding also involves arguing for the essential cognitive role of the citizens in constructing TA advice rather than only having participation in policy decisions based upon TA results. Actually, this essential cognitive role should be a double role. It should first be a regular utilisation in making TA of some descriptive, local knowledge the citizens have. This can help, among other issues, to make the TA investigation work more adequately and effectively. Second, an essential role for citizens' knowledge should be considered. Citizens may have a role in informing essential value choices for the cognitive process, in framing the cognitive process; what can be called a taking part in "cognition policy" in some general sense. (This requirement for participation in the framing process of expert cognition is different from the democratic political rights of the citizens of finally evaluating expert advice and making decisions.) It is important to notice that in neither of these roles is it claimed that the role of expert and citizen knowledge should simply be symmetrically ranked, even when they are both seen as essential. Together with their role in political choice concerning technological alternatives the two basic types of activities, forming and complementing expert cognition are the most important modes that citizens (should) have in informing technology policy.

Let me finish summarising the "moral" with two remarks. The first is that one has to differentiate between roles in principle and in the practical political arena where meaningful simplifications are not only preconditions for effective working but for its working at all. Hence, solving the problem of making a TA effectively workable through delegations or simplification of tasks is of the same importance as working on theoretical modelling of the TA process. Secondly, in some sort of political democracy, making TA just as constructing technologies should be seen as a sort of co-operative learning process. Two additional remarks are in order. With this emphasis on co-operative learning no ideal of moving toward even stronger consensus is claimed. Rather it is recognised that this is a process in which the dynamics sometimes lead to consensus, sometimes to clear limited dissensus only. A permissive democracy as a basic frame for unifying "Homo prospectus technologicus" with "Zoon politiken" should allow that the policing process essentially entails periodical opening up of any reached consensus, in the interest of reaching new ones through deliberating debates.

8.1
Remarks on the historical development of TA

As it is well-known, Technology Assessment was first institutionalised in 1972, being preceded by a six years preparatory period. It was introduced by the OTA as an additional advisorship to enrich the political decisionmakers (in the case of the OTA, the US Congress) with factual knowledge of the "side effects" of technologies so that the political decisionmakers could develop a more informed rational policy answer. This institutionalisation acknowledged the problematic lack of knowledge of the grave environmental effects of technologies. (This was the main target of cognition in the practice even when social impacts of technologies as topics of TA investigation were formally included into the early conceptualisation.) Policy makers and the political bureaucracy acknowledged that movements had some right to protest. To take regulatory action over from any movement the politicians accepted that some sort of institutionalised political regulation of the industry, to be realised by the state management administration, should develop.

The aim was to shape a technocratic and expertocratic solution to the cognition problem. Experts were called to widen their knowledge pools so as to understand the "real" problems behind the protests of the public. Even when it was acknowledged that the "lay" perception was of some real problems, the "calculative rationality" of experts was called for to step into the place of "emotionalism" and "lay perception", so that the solution should be a technological one. To use the well known term coined by Br. Wynne (1973) to characterize the institutionalised solution by the OTA, a "technological superfix" was aimed at so as to promote policymaking.

I want to emphasize some main characteristic features of this approach. First, technology was conceptualised as some sort of "thing", not co-produced with its "environment" and hence having "impacts" in a deterministic process. M. Akrich (1991) later called this way of thinking as thinking through "canonballs" and their effects.[1] Secondly, explaining this "thing", technology and its "effects", as they "really", "objectively" are, clearly seemed to be fully an issue of natural science, engineering (and economics). Moving inside this fixed demarcation, which defined the mode of perception of the issue, the investigation into the "effects" seemed to be clear: it needed exactly the extension of natural scientific and engineering (and economic) research.

Two main problems were born with this. They are from some point of view interrelated, but neither the problems nor their interrelation were recognised by those who made the first institutionalised effort in making TA. One problem was that regulatory knowledge was intended to develop without involving ethics (as "descriptivistic fallacy" somebody could term it), evaluation seemed to be a much more easy issue than the identification and analysis of the effects. The other problem was that, notwithstanding that it was intended in the interest of people, it was intended to develop a TA advice, without essentially involving the people into the process of developing it. This is a further main characteristic of the early way of making TA.

Early TA approaches were ambitious, both intending to provide a "comprehensive" knowledge of the "side-effects" and to realise an "early warning" capability. Both the claims for "comprehensiveness" and "early warning" have an important relation to our topic, the essential roles of participation

---

[1] An alternative perspective to investigation entails the following conceptual toolkit: processualisation, networking, permeable structures, co-production of "things" and their "effects", stabilisation, heterogeneous engineering, just to mention some elements.

of citizens in TA. I shall outline how both the task of developing some sort of "bounded rationality" instead of comprehensiveness (which is an easily misleading regulatory principle), and the "early warning" task may point to some essential role for citizens in the process of cognition.

Let me just touch upon some questions related to the task of early warning. In the early period of TA investigations, committed as it was to thinking in "things" and their "impacts", it was soon concluded that the task of early warning entails some paradox (The Collingridge dilemma 1980). This paradox seemed at first really fundamental. While I have no place here to show it, there is a large literature demonstrating that the paradox is much less serious than it seems if we do not commit ourselves to the "things have impacts" view. Instead I consider, however, another observation of robust importance. The identification of the task of TA investigations was conceptualised as an investigation with cognitive aims without asking if the requirements set for cognition would also put some requirements on social life. So it needed a long period, actually it was in the time when the idea of a scientistic TA was already heavily challenged by many, that Br. Wynne (1992) discussed a second type of paradox. This is that acquiring the knowledge needed to realise "early warning" as an element of the "upstream" regulation of environmental issues in society essentially brings with it some constrained changes in social life. The realisation of these constraints is precondition for "early warning" to be valid. Briefly, this observation by Wynne is about the necessity that social life must at least be some way standardized to be able to validly predict issues in it.[2] From the recognition by Wynne one has to conclude that not only technological development, but the cognition of it may call, in democratic societies, a legitimation through citizen participation.[3]

As mentioned, the task of doing TA clearly fell, in the early period, to experts recruited from natural sciences and engineering (as well as to economists). TA for environmental issues was identified as an essentially descriptive and analytical task with some, as it was claimed, need for a trivial evaluatory supplement. The ability to evaluate the identified impacts as "social actors", actually a task of ethics, seemed to belong to (natural or acquired through higher education?) capacities of these types of experts. Risk assessment ran through a similar development in which spectacular progress was made in developing the science of "objective risk", or as it is today more moderately termed: a science of quantitative risk. Fast real progress was achieved with this, positioning risk calculation right amongst the best sciences.[4] Un-

---

[2] Actually, this was a revival, and application to environmental assessment, of a recognition as old as that formulated by A. Comte, in the mid-nineteenth century.

[3] The requirement for separate gathering of different types of household waste, aiming at making its standardized treatment possible, is partly a requirement to provide for, as far as possible, standardized conditions for cognition.

[4] Quantitative risk calculation was mostly developed for nuclear risk issues. Risk calculation turned later to much more diffuse cases.

fortunately, another side was neglected simultaneously: that quantitative risk assessment is reductionistic. It was even evaluated to be its advantage that quantitative risk assessment takes only the measurable into account. But to be just quantitative, is neglecting important tasks compared to the complexities of the issues to regulate.

It should be emphasized that any reductionism in the cognitive approach when it is about socially relevant issues essentially produces winners and losers, and an interest in participation by the concerned involves the insistence on a different knowledge. This different knowledge about risks also includes qualitative indicators such as immediacy/latency, spatial distribution, distribution on human and non-human, voluntariness, to mention but a few. In short, when natural scientific experts were called to define the "real problems" and they developed expertise on a very narrow base, it is a small wonder that an enduring controversy emerged. "Denigration" of the "uneducated" was a weapon in the controversy on side of the experts, with growing distrust of the experts on the side of the "lay".[5]

The typical way scientific experts characterize the mode of how "lay" people comprehend risks was to mix up observation errors typically made by "lay" observers and the qualitative observations they also make, into one class, as "subjective", purely erroneous, or at least non significant observations. Admittedly, "lay" observers quite regularly produce warning evidence for committing erroneous observations, and of making fast or purely unfounded conclusions.[6] This evidence, from a scientistic, technocratic point of view, has been enough to conclude that the methodology of making TA should exclude the "lay": "devastating" shortages in cognition, regularly made by the "lay", could only confuse the serious, scientific, engineering investigation.[7] I intend to show that it is more realistic to accept a more complex view.

---

[5] According to the still prevalent argument "to make a difference between two risks having the same quantitative measure is but a cognitive mistake". It is important to observe that this conclusion is true in a decontextualized milieu and may prove false in contextualized ones.

[6] We know quite well that experts also often make "showy" conclusions of magnitudes of risks which are erroneous by magnitudes of order. (Shrader-Frechette (1993) brings a good list of such mistakes.) But this side of committing errors was not included into the early discussions.

[7] This bias is enduring. At my university, a professor, teaching "Electromagnetic environmental protection research", once met one type of a lay testimony and made it the motto of his lectures on the possible effects of some electromagnetic waves. According to this lay observation "Our neighbour has a new pacemaker and whenever he makes love with his wife, the garage-door automatically opens". The evaluation of the testimony is as follows: "it is so frappant that we anounce our course with this saying". Students inscribing to the course (that provides them with a very serious toolkit to investigate those electromagnetic waves), may learn from the motto that the best way is to escape "the" lay when a TA investigation should be made.

With this TA experts only applied the usual hierarchical demarcation in cognitive abilities, concerning roots going back in history as far as to the aristocratic differentiation between hubris and philosophy, and to the Enlightenment, its simplistic opposition of reason and sentiment. Having developed a characterization as the opposition between science and "pure opinion" in the 19th century, this hierarchical even exclusionistic demarcation became an enduring element in the self-identification of science and engineering. Contrasting this belief with the belief of the necessarily "distributed nature" of knowledge I shall try to develop a different view for TA.

I concentrate now on the early claim that experts alone are able to provide valid knowledge for TA. But life constrained to some practical compromise. Recognising the ever present "practical" problem of shortage of resources of most different types, as is always present in TA investigations, it sometimes became a practical compromise that even the "lay" may be involved in providing a substitute for sound scientific knowledge. ("Lay" knowledge would mostly mean having some knowledge of their own local conditions.)

I would like to make a short digression. Some need to moderate the original "rigorous" scientific claim for TA, not in relation to utilising citizen knowledge, was recognized even by those who were for this expertocratic model. I only remind the reader of the recognition Steve Weinberg made, as early as 1972, when he coined the term of "trans-scientific" for the case that important scientific questions for impact research can be scientifically formulated without the real possibility of answering them. His argument pointed to the lack of appropriate experimentation possibilities. One can of course add to this argument on limits to experimentation arguments about limits to calculability. And the TA task raised by the problem of the safe placing of nuclear waste can be an indicator that to most serious TA tasks can neither be a "cognitive fix" nor any but preliminary solution. The reason is that with extending the time period needed to assess, determined by the properties of the radiating materials, investigation for preconditions for safe placing becomes helpless.[8] It depends on basic perspectives on TA already if these pieces of recognition force a fundamental revision of the basic commitments and claims of the scientistic, expertocratic approach to TA and its role in the regulation process. The problem with these types of deficiences is that they may be decisively important in special, but socially rather important cases.

Following this early period of TA a different type of expertise, better to say expertise on a different level, was identified where the importance of ethics was brought into the foreground. It was interpreted, mainly in the USA, as the task of extending calculability to evaluation by utilising the utilitarian approach.[9] It was after including ethics as independent into TA, creating

---

[8] As an element of the complex problem one has to consider that geology cannot make realistic, reliable predictions for the needed long period.

[9] There is no place either to demonstrate the "innate" connection between a "liberalistic" type of democracy (the term taken from Habermas 1996) and the appeal

an independent evaluatory level, and then formulating the need of a third independent level, the rational management of technology, that a special type of TA, a rationalistic TA wholly based on expertise, got its fully fledged form. What earlier appeared on the scene as public and movement identification of and reaction to adverse environmental impacts of technologies had become completely "rationalised" and received a full-fledged expertocratic overtake.

Insisting on this framing, a further, complementary task has been introduced, that of the enlightenment of concerned people. This enlightenment however was in an expertocratic vein, envisaged as a strictly one way communication, i. e. to help the concerned people know what their problems "really" were and how they "really" should be solved. Taking these concerned people into account as "lay" persons, that is, persons with a lack of scientific, engineering, social science, humanities knowledge and research capacity, meant that a one way hierarchical communication, communicating of the experts' results to the "lay" citizens was to add to the expertocratic TA.[10]

One could be inclined to wonder if a wider range of experts were to be added to the list of experts to take part in TA investigations. At least to my knowledge no experts of aesthetics have been invited to make TA knowledge fuller. The only exception has been the inclusion of some psychologists on the reflection level. These psychologists took for granted the "objective knowledge" of risks and tried to identify the "lay misperceptions" of the "real issues". They concentrated on describing and explaining all the shortages everyday "lay" people showed when they tried to understand the risks around them that were caused by technological development. It is true that everyday people are clearly "lay persons", repeatedly making all the mistakes too, in the cognitive process that one really makes when s/he is not a scientist, engineer or psychologist, i. e. trained to avoid some sorts of cognitive traps. Their real mistakes were identified by experts and politicians as "subjective opinions" against "objective knowledge",[11] and hence were evaluated as barriers to the rational solution of problems of technology tackled by expertocratic TA. But this characterisation was distorting because it is one-sided. It is significant that under "subjective opinions" massive illiteracy in science or engineering were put together in the same basket with value commitments influencing risk evaluations by the "lay".

One can say that two main funding demarcations defined how expertocratic TA and its relationship to "lay" citizens became stabilised. One was the analytically sharp contrast made between knowing, arguing and decid-

---

to utilitarian calculus, or for demonstrating the shortages following from applying a utilitarian approach. Yet, it is important to indicate the connection and the problems.

[10] It is characteristic that the "lay" was defined in purely negative terms.

[11] Which was simply valued more highly as real, trustworthy, and legitimate. The scientistic, expertocratic modelling of the real situation was simply accepted as unquestioned basis to evaluate "lay" perception.

ing on a purely power base, or in a finer form, deciding through bargaining. The other contrast was made between ideal situations and the constraints of poor practice. Fully in the tradition of the Enlightenment, or, better, in the tradition of its dull, positivistic formulation, scientific cognition was contrasted to irrationality, pure power, or "subjective" values and preferences. The aim was then to make decisionmaking scientifically rationally based, that means to provide the decisionmakers with the best scientific knowledge of the impacts of technologies as to avoid "irrationality" in the knowledge base.[12]

The other demarcation caused less problematic tasks for those introducing it. One had simply to accept that making Technology Assessment is an immediately practical, goal oriented research and in this way will be always heavily constrained by practical shortages. For our problem it followed that TA evolved a special type of "tinkering" nature. That means that some knowledge was accepted as coming from non-scientific, also from "lay" resources, from everyday citizens.[13] A sort of "best practice" has been realised in this way several times.

Until recently a presentation arguing for the involvement of "lay" persons in the cognitive processes of TA would have been sufficiently framed by these considerations. The tasks seemed to be clear and divided in two parts. In arguing for "involvement" of the "lay" one has to show the perhaps various roles "lay" persons can play in producing "strategic" knowledge by developing and utilising "local knowledge", useful for making a TA. Local people may know something from immediate experience about issues experts have not had access to so far, or not had time, or interest to investigate. Besides dealing with this possibility, an earlier presentation would have also dealt with the basic problem concerning if the "lay" should be involved in evaluating long run perspectives of technoscience, a capacity that has been much more inimically welcomed by many experts, industrialists and politicians than the utilisation of local knowledge. In this case value commitments of the citizens become interesting, and they would be called for participation in order to make evaluative decisions on the readymade alternatives prepared by experts. But one can ask if the cognitive processes themselves in which these alternatives are worked out should only be framed by the experts, preserving in this way their cognitive isolation, or if the citizens could also be meaningfully involved, perhaps even essentially involved in this framing process of cognition. Some developments seem to change the story, and, it seems, bring a need for profound reconsideration of the "lay" participation in TA

---

[12] We already saw that this ambition led to define the qualitative as "purely subjective", i. e. not formalisable. There is no place here to deal with newer developments, successful approaches to represent qualitative judgements by computers are already on the way. Computers provide aid in making qualitative judgements more formally treatable as we have been used to thinking. With this I do not want to claim that they can be fully taken over.

[13] The process is really a tinkering, i. e. taking the resources where ever they are found and, if urgent, in the form they are available.

processes in this relation too. Considerations about the limits of rationalisation of uncertainty by developing risk calculations as well as the need to consider qualitative aspects of risks lead to a rethinking of the hidden rationalistic presumptions that would still hinder a more profound participation. I shall come to this point later to explicate it.

The typical problem between the experts and the "lay" in the early period of TA investigations was not participation by the everyday citizens in TA investigation processes but the "one way communication" process. The emphasis was on raising the knowledge of the citizens about the adverse effects caused by technologies around them, as they were identified and analysed by the experts. This "one way communication" entailed a most important task. New technologies often caused effects, the identification of which left the everyday observers in complete ignorance about their nature (people observed cancer cases without a real capacity of connecting them to their possible causes, etc.). To make the "lay" conscious of the dangers became a task that many times and in many ways really needed enlightening efforts by experts. This is a task that should be continued when introducing new technologies.

A further step, after the enlightening efforts, is to utilise the "local" knowledge of citizens to make TA investigations and technological construction work more effective. Recent debate over citizen participation has been less with those who were ready to exclude the "lay" from any participation in the cognitive processes because the usefulness of "local knowledge" in the above mentioned, instrumental, practical sense has been proven enough. One has further to reconsider that with any further development of science and technology the common interface, at least with everyday life will be reproduced. Lay involvement also satisfies certain democratic political needs, utilises their legitimating force through identification with appropriate technological plans when the citizens co-operate in technologically reconstructing their "local", everyday life.

The question still arises if people could cognitively participate in formulating and solving overarching problems of science and technology policy, just as in other public policies.[14] The fact that citizens more and more vindicate the right to participate in such evaluations does not change the situation that their participation, by some considerations, may not be proven essential or at least effective. One can say that representative democracies have developed also because immediate participation in policy processes is severely restricted. And to participate essentially may be hindered by cognitive reasons. Of course, some cognitive barrier may be decisive; "illiterates" concerning the relevant knowledge exclude themselves from any dialogue. But a debate about citizen participation in science and technology policy decisions, like regulation of genetically modified food production or xenotransplantation research is not only heated because too many "illiterates" insist on taking part in it. It is about social value choices, about long run perspectives on how

---

[14] The fact that they actually do it in various forms does not justify it.

people intend to live, and the debate is also inextricably interwoven with the forms of democracy accepted. Should dealing with problems of these types be left for representatives only or be pushed toward more participation? Recently, for different reasons, including distrust in experts, there is a growing demand by citizens toward more participatory forms of TA and constructing technology.[15] On the one hand defenders of expertocracy (and representative democracy) seem to emphasize the cognitive difficulties, "lay" citizens certainly have to acquire as appropriate factual knowledge about technologies in the forefront of research to participate effectively. On the other side defenders of participation put emphasis on their essential role in participation, either only for political considerations or, I shall come back to this question, for reasons based on the peculiar, cognitive sociological structure of research.

Let me turn now to another side of the complex relation between experts and the "lay", the citizens. The experts were identified as they were found in the practice of TA and technology development at large. The term "experts" meant natural scientists, engineers (and economists). Until now the problem was spoken about if people would be able and acceptable to participate in developing technoscience and its policing in relation to what was needed in natural scientific and engineering knowledge. But to interpret the importance of new technologies in their lives, people need to have some sociological, phenomenological, hermeneutic, aesthetic knowledge too. The claim by some radicals and not only by them, is often that citizens are in this regard "experts of their own situation".[16]

I think this evaluation is seriously misleading. It is confusing in that people necessarily make their own life through observations of and reflections on their experience but also their life is interwoven by science and technology and is itself a topic of social scientific and humanities investigation. To insist on the idea that "citizens are experts of their own situation" could only be hold by a radically anti-science attitude (science in the meaning of social sciences and humanities).

I turn now to a point that is more than trivial in democratic societies, but is nevertheless often not taken into account, at least with appropriate weight. It is that different groups of people in a complicated social place may differ-

---

[15] In developing participatory forms of investigating into and evaluating long run technoscientific and technological developments, what I am inclined to argue for, certainly brings with itself immense practical problems of knowledge communication. One only has to look to the new "labelling" requirement for food in the EU, where to simultaneously provide the simple consumer with rigorous, comprehensive and simply understandable written information on their food on the food packagings, relevant for the most different types of consumers, to understand that the democratising process of regulation can move itself to desperate situations.

[16] Another claim, a rightful claim in my view, is also often to be found in critical literature. According to this claim natural science and engineer experts are not experts of the local life problems of citizens either.

ently perceive what happens to them as the "official" expert TA reports on it.[17] Having identified that the "official" expert reports decontextualize the issues at stake, or contextualize them differently, and that this decontextualization, a different contextualization damages their interests it is only natural that a search for alternative interpretation starts.[18] Actually one can take as typical that technological controversies over the values included in technologies, and the contexts technologies would be applied, hence their "effects", emerge when citizens or certain groups of them recognize that their expectations about life are also at stake with technology political decisions. And it is also typical for technological controversies that as many alternative assessments can be made as there are relevant interest groups, including perhaps different groups of citizens.

The simplifying interpretation of technological controversies identifies these controversies as if they were between two parties, "the" industrialists and "the" citizens. One can be inclined to evaluate this situation in the sense, that committed to expertocracy, citizens by developing alternatives, because of lack of appropriate knowledge, just commit mistakes, a certainly typical case. But to forget about the other side, that evaluations by citizens may contrast with the expert reports because they use different values is also a typical case.

Looking from a different perspective, that means differently framing the "same situation", a different knowledge of the same issue can be produced, also as alternative expertise. It is typical for the scientistic TA that it has been claimed that, in principle, for the same issues the same expert results should be obtained. ("Narrow definition" of the so called "experts" dilemma, based on strict methodologism.) With this claim this methodology generalized some very limited possibilities where strong enough restricting preconditions were set. For, this methodological rigour cannot be realized in most of the cases because of the principal freedom in choice of the cognitive values that are necessarily needed to orientate the investigation.[19] (And these cognitive choices may differently serve for social value commitments.) Of course, investigations

---

[17] A striking example of the many-many is the case of the reports on a dam built up in Bangladesh. According to the report by the builders it was a full success, measured e. g. in terms of energy production, security against floods, etc. The UN that provided for financial support had in astonishment to recognize that some (layers) of citizens perceived the same case differently, e. g. as lost of their living conditions.

[18] R. Sclove (1995) summarizes and comments on the Mackenzie Valley Pipeline story as a case for many when the residents quite differently identified the changes to come after setting technologies than the official evaluators. This case clearly shows that because of indifference, lack of knowledge of local conditions, having a perspective different from the residents, as e. g. a narrow industrialist perspective, led to Technology Assessments that deeply hurt interests of the residents.

[19] This is the problem of necessary "valuation" already recognized long ago by E. Mullin (1973), among others.

can be standardized. Many arguments for standardization efforts in scientific methodology and then for the methodology of regulatory knowledge production can be specified. (One can ascribe a status of definition, a normative force to standardization efforts.)

Concerning its relevance to the types of expertise in TA at least one claim certainly can be made. Some sort of phenomenological, hermeneutic knowledge to be able to interpret changes in residents' lives should belong to the armoury of TA. Any group of TA experts, recruited only from natural scientists, engineers, economists rarely can be requested to have it. Another, stronger claim may also be raised. It is, that citizens quite regularly conceptualize their social situation differently from the conceptualizations the TA experts or technology developers inscribe into them, even on the descriptive level.[20] Describing "social reality" requires knowledge different from the classical expertocratic TA based on a narrow knowledge basis. It requires to recognize that different descriptions of the same social situation due to technological changes may equally be "valid", according to the different value commitments the descriptions take as their framing. Any decontextualizing methodological norm setting may call for protest by those whose life situation may be inadequately conceptualized. Provided they recognize the problem, I will put the earlier statement, for I shall argue for the essential mediator role of experts in social sciences and humanities, committing themselves to acquire the value commitment structure of some groups, including the citizens. Their essential advisor role is often forgotten about by those who argue for lay participation.

All this is rather clear with risk research, that is becoming one of the most important research fields. Value choices frame any scientific research, so any risk assessment. If the essential role of value commitments in risk research is clearly recognised, then it becomes understandable that to stick only to quantitative risk assessment in a scientific characterization of risky situations is not to stick to objectivity, in the sense of unbiased knowledge,[21] but to commit to a peculiar reductionism framed by a specific, reductionistic

---

[20] Making a description of the social situation involves framing of the cognitive activity by choosing for cognitive values. Fixing the object of the investigation, the chosen investigation techniques, the evaluation criteria of the investigation process and the knowledge produced, in short the whole cognitive process may be consciously or unconsciously influenced by cognitive value choices so that different specifica of the investigated issues, essentially important for the social practice get emphasis or will be fully abandoned.

[21] This does not mean that a quantitative risk measure may not be real. If the criteria of reproducibility, etc., provided for by the scientific analysis are satisfied, quantitative risk analyses may lead to real, "hard" knowledge. But it is also sure that not only the qualitative criteria of risky situations are systematically excluded from quantitative risk analyses but it is also true that some "quasi-scientific" quantitative risk characterizations appear where basic preconditions for quantitative risk assessment are missing.

value orientation. This reductionistic value orientation excludes qualitative values from risk considerations, it plays them down as purely "subjective".

What is at stake with this is firstly to acknowledge that risks have more than only the characteristic measures that prove generalisable (probability and quantitative damage), that there are many other less general, qualitative ones. To reintroduce them into a less restricting scientific treatment of risky issues and make a choice between them, together with quantitative assessment of risks, is interconnected with choosing for some types of life. In the same way it is to understand that committing to quantitative risk characterization only is not only a decision for cognition, valid for everybody, but a decision for the type of cognition essential to some sort of life. If somebody would acknowledge this point, (s)he could still argue that it is about how to connect "objective science", that means quantitative risk assessment with life, with the choices different people may differently make. That means (s)he would identify the situation as overcoming the "underdeterminatedness" of social decisionmaking by science. But a further claim is by now crystallized by some group of researchers. That is that the descriptive cognitive problem in risk studies is already always "underdeterminated" without choices of what sorts of qualitative characteristics of risks should be included in the investigations. These choices are to be made as preconditions to enable the scientific approach to work.[22] If value choices are at stake with cognitive problems[23] there may be an inclination to claim that this is a point where immediately involving the "lay" instead of experts is essential.

I would like to make some comments on this view raised by the author collective of an important recent paper intending to reorientate scientific approaches to risk assessment.[24] The mentioned conclusion is drawn in the following way. To be able to make scientific risk assessment one needs decisions on scope (choice from the multidimensionality of the features of risky situations), incommensurability and ignorance as crucial framing assumptions. These decisions have a fundamental status indeed, in the sense that no risk assessment can work without conscious or hidden, previous decision on these problems. The authors conclude that this leads to an argument for greater public involvement to realise the essential decisions so as to enable the investigation to work. They claim that in this way the "active stakeholder engagement in the appraisal process becomes a matter of analytical rigour".[25] I accept that without systematic reference to the values, priorities and scope displayed in wider discourse the crucial framing assumptions adopted in ap-

---

[22] From this point of view to reduce risk assessment to quantitative risk assessment only is a degenerated approach which proves decisionistic in choosing for scientific method.

[23] Immediately as choices between possible alternative cognitive attitudes and substantial value choices needed to make the cognitive orientation well-defined.

[24] Actually, this is the claim the authors of an ESTO Project Report (Andrew Stirling 1999).

[25] Stirling (1999), p. 30.

praisal cannot be validated, but I think that it does not make immediate participation necessarily. The needed validation for making risk assessment work as a cognitive process may be also made as expertise, different from those already included into the cognitive processes of risk research. It is arguable as I shall specify that participation will only be indispensable when a type of political decision about a form of democracy is also made. More than that, different forms of citizen participation is possible to satisfy the same essential cognitive need, mentioned above. That one that immediately offers itself as accepting the value commitments of people as they are seems even not the best for some sorts of political purposes, just for those purposes aiming at securing the best position for people in a "knowledge society".

As it is well known two main strands coincided to produce an essentially different political and cognitive situation now in which problems of modern TA are identified. The one is the strong requirement by citizens to co-define through participation, to become immediately engaged in the political processes in general in the developed democracies, which includes involvement in different types of public policies. The other, in my opinion much less weighty strand of changing attitude that leads away from expertocratic policymaking in technology, is the recognition that the "pure" expertise is not mighty enough to solve the very expertocratically defined problem of TA investigations; that the process proves to be too complex for this simplifying approach, provided expertise is only recruited from mathematicians, natural scientists and engineers. Several questions raise then, as to how the political requirement of public participation should be realized, which types are urged by different understandings of what a democratic policy is and should be, what types of cognitive place(s) are to be ascribed to public participation, and how the "lay" should be educated to be able to make an informed assessment, including both the possibility of informed consent and dissent and how the experts also should be educated to become able to co-operate. I come back to this point after having dealt with different forms of democracy and the cognitive characteristics of lay knowledge.

8.2
TA as a democratic requirement, types of democracy and "technological citizenship"

As already mentioned, introducing institutionalized TA in the early seventies was a compromise in democratizing technology policy. It took the task of getting some rights of citizens acknowledged over from social movements and yet essentially it did not include the citizens in the process. TA was introduced in a way that fitted a "liberal" form of democracy. (It is a simplifying but useful classification of the basic (ideal) types of democracy identifying them as a "liberal" and a "republican" or "communitarian" form, taken these terms from Habermas (1996).) As Habermas identifies these idealtypes the first is a representative mode of governing that puts emphasis on negative rights and privacy, with the second being a mode that emphasizes the active participation of the citizens in governing their public life. TA, as any other policy advice may move between two (ideal)types, reflecting and enforcing the policy conceptions. Taking the position of "representative democracy", an appeal to bureaucratization, hence to formal rationality and expertise as delegation may be identified as appropriate. Habermas identifies a third idealtyp too. This is, as he calls it, the "deliberative form". This form tries to overcome the weaknesses of the other two idealtypes, unifying their strength, meanwhile avoiding their weakness.

Recent institutionalized forms of TA are clearly part of a "liberal" form of realizing democracy. To construct a different type of democracy requiring as much public sphere creation for self-governing action and community creation through public dialogue as possible require a changed institutional form and content of TA, too. It is useful to introduce here a differentiation between social accountability and immediate public accountability. One can approach the differences of these idealtypes of democracies from the viewpoint of how social accountability is related to immediate public accountability. There is a strong pressure by citizens to raise public involvement in TA. This claim can be satisfied by a move toward more "deliberative" forms or by turning over to "communitarian" modes. Both these forms presuppose some sort of activity that can be termed as realising "technological citizenship".

"Technological citizenship" must stand in the centre of considerations for those who want to push political decision-making toward more participation, even when the criteria of technological citizenship, in a weakened form, can be formulated for pure "liberalistic" democracies too. "Technological citizenship", to briefly formulate it, is a form of citizenship in issues of technology policy. Before going into some detail of what "technological citizenship" should mean, it is useful firstly to summarize some facts about the "effects of technology on life" that make a participatory form of citizenship important for issues of developing and regulating technologies.

The American philosopher of technology, Langdon Winner stated in 1979 the challenging and mobilizing that technologies are not tools to live with, but "we live them". Of course, from one perspective it seems true for the whole of history that technologies have always been more than simple tools.[26] What is correct and important in the statement, I think, is that the pace of technology is immensely accelerated now, so that technologization is much more pervasive than it was in any time earlier. Such spheres or/and on such levels are included into technological reproduction as they have never been involved. Initiating changes in life through technological evolution has become a central

---

[26] Even if one is not ready to follow the technology deterministic approach used by Lynn White, Jr., one can accept that the stirrup and some other "tools" had an essential role in the development of feudalism, that they were essential ingredients in the complex process in which life changed into a new form.

issue for society. Technological innovation is in many respects a consciously undertaken social experiment with life forms, forms of connection in which the broader range of consequences is only foreseeable in a rather limited way. So it is a serious challenge for public assessment not only for assessment by the researchers themselves. It is a small wonder that the reaction moves between exaggeration and despair, with critical trust in the middle.

In this accelerated pace of technological change (I am inclined to accept the "third technological revolution" as an appropriate term (Thurow 1999)) hidden and open models of life serving to develop and utilize new technologies as well as the unintended effects become especially important. "Lay" preconceptions about life, about consumers' needs, about possible future consumers in the head of developers and producers, the often lack of time and space to make trials with new technologies by their users (even by their developers), make many users concerned about the values[27] of new technologies. These concerns may multiply the importance of the fact that technologies are not only used as tools but "we live technologies", we construct our society and lives technologically.

Technologies are polipotent, functioning in focal and non-focal ways. Their developers are necessarily "abstract" concerning their ability to foresee the practically unlimited application situations.[28] People experiencing them in context several times are unable to identify the "effects" correctly. Technologies are not neutral in their "effects" on life, they function as "technological experimenting on and with society". Human actors get their social identity largely through technologies, experimenting with them just as technologies (actually their developers) experiment on them. Hence the obvious, very natural need for self-determination by the citizens exists and as a part of it participation in governance of technology is needed. All this requires, in my view, reaching a level of "technological citizenship" for both (!) the "lay" citizens and the "abstract" experts.

---

[27] That "lay" engineers, "lay" in the social requirements for functioning their technology, are still much too numerous as one can exemplify in many cases when different cultures meet in technology transfer. According to an Egyptian PhD student of mine Western architects, supported by an aid programme, constructed blocks of house with big windows and with wide streets among the houses, neglecting both natural, geographical and social preconditions for usability. On the other side just knowledge management efforts show some important tendencies that individualistically shaped approapriations of technologies can also be developed through processes of fine-tuned interactions between producers and consumers. Certainly, the process of "user involvement" into processes of technology development by the industry is a recognition of this problem and an effort to eliminate it.

[28] Charles Perrow's famous warning concerning the essentially limited predictability of the behaviour of technologies because of the complexity and differences in the situations they may be involved was just a first recognition of this, in relation to dangers. See Perrow (1984).

Frankenberg (1992) introduced the term "technological citizenship" into the literature with the aim of producing an overall normative framework by which basic problems of governance of technology can be expressed. To have technological citizenship is to understand together with a "social contract" for technological society. A "social contract" regulates and re-regulates as a political regulator who controls, by what rights, and how control is realised. A need for an informal contract in issues of technological development is quite obvious nowadays. It needs to be about how to share responsibilities in a society that commits itself both to innovation and security. The aim of Frankenberg is to provide for a model of politics which "reconciles freedom to innovate and the affirmation of the autonomy and dignity of laypersons and the assimilation of laypersons with their world."[29] One can say, following Frankenberg, that citizenship entails equality of membership, participation, and status or standing of individuals within a bounded realm governed by a state. This membership is woven of rights and obligations that are guaranteed and enforced by this state and that aid in the achievement of some overarching goal. This goal is normatively valued by reference to a set of more basic values. As Frankenberg puts it autonomy, dignity, and what he calls "assimilation" are the most important of these values.[30] "(T)hey are to guarantee humans that their autonomy and dignity or respect for their inherent worth and for that of their agency for self-direction will be protected as of right." Technological citizenship symmetrizes experts and "lay" persons in relation to their political rights concerning technological issues.

It can be seen that the starting point of Frankenberg is the "adversary model". He conceptualizes two main actors on the technological political scene: the experts and the citizens, in some general sense. They are identified as being in a struggle with each other. The trivial starting point for this model is that experts have a right to do their work because they have the status and the knowledge they need to fulfil it. "Technological citizenship" (TC) is first of all a status in the political structure of society. It is defined as "equal membership", participation, and standing or status of persons as agents and subjects within a realm of common impacts to at least one "technology" – or instance of consciously amplified human capacity – under a definiable state that governs this technology and its impacts. Such status is defined by a set of binding, equal rights and obligations that are intended to reconcile technology's unlimited potentials for human benefit and ennoblement with its unlimited potentials for human injury, tyrannization, and degradation. Such status, rights, and obligations are thus intended to reconcile democracy for key subjects of technology's impacts with the rights of innovators to innovate. TC reconciles the autonomy of laypersons with the autonomy of experts."[31] But this status is not ascribed to persons without any further requirement.

---

[29] Frankenberg (1992), p 462.
[30] With "assimilation" he coins a term opposite to alineation.
[31] Frankenberg (1992), p 462.

The public must raise to the level of "technological citizenship" to be able to participate in the process of the governance of technologies. Those who are accepted as "technological citizens" should be capable of informed consent and autonomous thought in issues of technology – society relations, including the relations of technology to the environment. Frankenberg understands autonomy in a way what can be termed as "communitarian liberalism".

"Technological citizenship" entails, among other duties, obligations to learn and use knowledge relevant to governance of technology. Frankenberg, moving inside the model where the perspective is on getting under controlling technologies by experts and the citizens, puts emphasis on the requirements that they should know what the hazards are, how hazardous they are, how they operate, how they are monitored, why they exist, and what the alternative means to the ends they produce are. They have to know the synergistics, cummulativeness of the effects, and the main possibilities of monitoring and controlling them. Having acquired this knowledge level a valid public acceptance, deep and informed, subjective, voluntary, rigorous, and rich can be developed. Beside these criteria for technological civic literacy and civic virtues (critical trust), a critical judgemental capacity is also essential.

One remark is already much in order. This is that the knowledge requirements for reaching the status of "technological citizenship" are rather strong, but deliberative. On the one side they neutralize the usual objection based on general "lay" illiteracy, measured by some methods that are simply copying requirements for expertise and weaken them. Citizens should not be experts but they should acquire the knowledge that enables them to "orientate" themselves. On the other side this requires very serious, repeated efforts to cope with the pace of changes. Among many other things this task of "orientation" requires acquiring an appropriate knowledge base for the citizens for the solution of everyday problems of how to become an informed user able to exploit the different capacities of new technologies. To not speak about participation with equal political rights in technology policy this is already a common and serious problem. Life world knowledge without repeated learning in any form is certainly not enough to enable involvement in long range technology policy issues. It is problematic in everyday life accommodation processes already.[32]

But people may insist for active participation. (Reasons based on the obvious facts of the continuing environmental deterioration count to these reasons.) The emerging contradiction can be solved only if it is possible to

---

[32] I hold for one of the myths around "lifeworld" that people living in their "lifeworld" acquire some capacity of assessing the issues "colonizing" them. W. Krohn details an exemplary case to show that persons in everyday life get into dilemmas when confronted with the possibilities of new technologies. It can be taken as the general case. Lay people have to acquire the ability to utilize new technologies for themselves. See Krohn (1997).

acquire appropriate cognitive capacity and knowledge without necessarily becoming an expert.

Before I turn to this problem I just would like to mention that in my understanding "technological citizenship" puts burdens on the experts, too. Taking into account that expertise (and I speak now about experts trained in natural sciences and engineering) leads to new technologies, to new uses of already existing technologies and that these activities have the nature of "social experimenting" through their inclusion in the lifeprocesses of society, experts must also be required to acquire appropriate knowledge to be able to understand and estimate the social consequences of their activities, at least on the same level as it is required by "technological citizenship". The criteria are symmetric in this relation. While getting involved in the process of developing and using technologies experts may be insulated in their roles when it is defined only in an abstract way.

Let me make here a short remark. An important recognition by the social constructivistic reconstructions of history of technology is that technologies are always somehow "in the making". That means the possibility to actively participate in forming technologies is always given through the whole process from planning, developing and using technologies even when not on the same manner and deepness.

8.3
Cognitive preconditions to develop and practice "technological citizenship"

One can approach questions of cognitive preconditions from a bad perspective. It is the scientistic model. It insists that scientists, experts in a broader sense, have the capacity to know something but the "lay" have not. The lay is defined negatively. In its caricatured form, there are no graduations involved, you are either expert or you are the opposite. It insists on the necessarily one way communication with the "lay persons", i. e. looking at the lay persons as simply scientifically not ripe persons, unable to any dialogue for their "illiteracy".

At this point I go back to the often emphasized statement that the "lay" are scientifically, technologically illiterate. In one sense it is certainly true, (Durant (1988) brings a rather comprehensive survey in this relation), just as it is true that often experts of one field are illiterates in any other. The question is however, how far this type (!) of illiteracy partly measured in non-important pieces of information from the point of view of technology policy prevents the "lay" from participating in TA and technological political decisionmaking. Another question is if some techniques can be found to enable the citizens to recognise and assess the relevant alternatives of a technology policy over (perhaps future) technologies, the details of which, even in many essential features, they remain illiterate about. (One has not to forget either that measures of knowledge of lay people are context dependent, and may involve hidden or even open political purposes.)[33]

(I make a short excursion to wonder a bit on the social role of some ignorance by people. Everyday citizens are often observed to not being interested even in the relevant knowledge of e. g. dangerous technologies that are clearly necessary to understand his/her basic life-conditions. This is true, and, unfortunately, it deprives these citizens of being "technological citizens". One can wonder that instead of raising participatory requirements why many follow this way. The answer seems to be developed through a choice of one of two patterns. One explains the case by emphasizing the positive role of ignorance in survival, somehow as a modern incarnation of the famous passivity of many people in India. The other may be an argument for a "representative democracy", that many are inclined to define themselves as "lay" having trust in those who represent them.)

Coming back to the knowledge potentials of "lay" people, I repeat my claim that the correct question is not concerning the participation capabilities, how much less the "lay" know than the "experts". The correct question is what the relevant knowledge is for "lay" people to reasonably participate and how they may acquire it, provided that the true horror stories are neither decisive arguments nor negligible. The nature of the knowledge essential to participate in constructing technologies, or in technology political decisions is either of a "strategic knowledge" nature or of a phenomenological, hermeneutical, or ethical one. Let me take first the issues related to gaining "strategic knowledge", to help the effectivity of action. The conclusion will be that "lay" knowledge can serve as a complementary and falsificatory control for expert claims.

It is interesting to investigate some famous mistakes made by scientists because this can lead to some systematic knowledge about where the knowledge of everyday citizens can be a useful complementing element or corrective to expert knowledge. This is not about "denigrating" the experts but rather the aim is to show some systematic points where they interact with "lay" people. Let us take first one historical case, that of Liebig and his use of potassium as an artificial fertiliser. It is rather well known that Liebig went to a lot of efforts to produce an insoluble salt of potassium and offered it as fertiliser. When farmers insisted on reporting their failure to utilize it he interpreted the situation from the standpoint of an exaggerated scientistic ideology and damned these people for their failure, until he, after twenty years or so, recognized his own, actually quite obvious mistake. It was that the plants were unable to take up insoluble potassium compounds.

To get to a less biased evaluation of the capacities of all the actors in developing, using and policing technologies it is nevertheless important to see that scientist and engineer experts systematically make mistakes:

[33] One has to remember in this relation the role of quantitative IQ investigations in 60s and 70s in the States as exemplary case. See detailed in Fischer (1992).

a) According to their own criteria.

b) Sometimes constraints on modelling are so strong that they are only able to produce a limited type of knowledge, only, leading to obviously false practical advices (the argument from the "educated idiot").

c) And one should not forget about the temptation of getting misled through a scientistic ideology. According to this scientists usually claim at least in principle a validity even when the rude reality does not allow the development of any scientific solution to the problem at issue.

A series of mistakes made by scientists as striking examples are easy to gather. Wynne (1992) reports on the Cumbrian radioactive fallout case where false predictions of the measure of the declining of radioactivity of Cesium were made. The mistake made was rather typical: measurements under laboratory conditions were transferred to fields, where the preconditions were different. Using the Darcy equations for calculating cases of leaking fluids under real geological conditions is several times not more than rhetoric, because the measured values are simply too wrong.[34] It is not enough, of course, just to demonstrate mistakes by scientists. In arguing for a role for "lay" observers it is questionable if some of these mistakes are open to detection by "lay" observers.

There are a lot of cases when "lay" observations led to the recognition of new research problems. According to records, schoolchildren in Minnesota were the first ones who recognised deformation of frogs.[35] It was identified by experts that the deformations were effects of environmental pollution. The Woburn story is widely known, especially because it was filmed some years ago. It is about housewives who recognised the unusual frequency of leuchemia cases in their neighbourhood. Everyday people reminded Carson of the birds, dying from DDT. All these cases were outside the attention of professional research earlier. One can risk the conclusion that systematic "lay" interest can usefully complement professional research, both in the role of additional information and as possible falsifiers. Once again, we come up against the general case of distributed knowledge. First, there is, both a qualitative (as a whole) and an only quantitative difference between expert knowledge and the knowledge of the "lays". Lay people also make experiments, draw conclusions, make some statistics, look for sensible indicators.[36]

Science has a long time ago differentiated itself from non-science. "Lay", better to say amateur research, has continuously accompanied scientific research, especially where the field research has been typical, like geology, or

---

[34] Shrader Frechette (1993) brings a huge amount of these types of methodological mistakes by the researchers at the investigations at the Yucca Mountain.

[35] The coming list of cases is mostly taken over from Sclove (1995).

[36] The public opinion in Hungary insists on believing that suicides committed by some police officers in Hungary, maybe related to the working of an oil "maffia" in a small county, may be unusually numerous. Recent inquests seem to approach this original opinion by the "lay".

botany. "Lay" research could still lead to important supplementary knowledge in these fields. Tasks, such as developing GIS[37] or realizing comprehensive measures of environmental protection show some important cases.[38] It may be that no detailed scientific knowledge is available at some places and local knowledge must be utilized, including that of indigenous people. Consumers of different sorts of scientific or technological services nowadays make self organizing efforts to systematically develop a parallel knowledge base, partly to control the "official" knowledge on the field. An example is that of the "patients' records" in relation to knowledge of medical doctors.[39] Clearly, a type of "lay" variant of the most aristocratic type of expertise has been born and has developed rapidly with the help of and by experimenting with connections through emailing. The consequences may be rather important and would call an army of analysts to observe them.

All this has been about natural scientific knowledge. One has to look at the social sciences, too. Engineers working on their construction many times forget that they are simply laypersons in issues of social consequences of engineering constructions. In engineering work it is sometimes presumed that they make rational assumptions about the possible users. It is interesting that some analysts counterpose the everyday citizens to this as if they were experts of their own situation. The observation correctly is, in my interpretation, about the trivial capacities of lay persons when accommodating something in their well known everyday working or life milieu, something that can be proved by co-operation of experts and the "lay" either true or false.

Science consciously, and any expertise perhaps only unconsciously realizes a decontextualising (modelling) endeavour. It is but trivial that in the manyfold ways contextualized milieu of everyday life citizens have to take over some sort of work as providers for complementary, or falsificatory observations if the model, as is typical proves, to be too simple or runs against the preconditions.

---

[37] Geographic Information System.

[38] The funniest case of including "lay" people's knowledge into engineering construction work I met with is in an inquiry into the "wisdom" of people in history, to learn by the help of archeologists, where their houses were built to gather knowledge of the move of the sea at a seashore to use this knowledge for dam building plans.

[39] As Akrich and Meadel put it in their summary at the 4S conference 2000, /Discussion lists/ allow, at an unknown level, exchanges of various forms of experiences (what is to be ill, how to deal with treatments, how to interpret various sensations, symptoms, feelings, what to expect from physicians etc.) and also exchanges of medical and scientific information. This phenomenon raises several questions: does it lead to the constitution of a new common knowledge and thus to the emergence of a new form of patient expertise? Does it modify the relationship to medical professionals? Will it result in the emergence of new collectives of patients? 4/EASST Conference, 2000, Abstract book, p 2.

I mentioned the Cumbrian case and other cases as telling examples of trivial falsificatory capacity. For a generalizing role as alternative cognition doing popular medical research, patients' lists may serve as examples. Their peculiarity is from an epistemological point of view that perpetuate, either forms of cognition from pre-scientific times, a still very important method of cognition in many places in the third world, or continue practical empirical investigations, several times reinforced with new tools, like using computers. TA, always in shortage of resources (lack of scientific knowledge on the field, time for investigation etc.), may usefully integrate these types of knowledge, when critically selected.

A peculiar mixture of expertise and "lay" participation may occur in so called community based research where local communities set the research agenda. The goal is here to immediately couple research with community needs.

Constructive Technology Assessment developed an important aid to help citizens to actively participate in knowledge production for TA or getting involved in alternative styles of technology development. The method of inviting experts who contradict each other has been very helpful for the citizens to understand the existing alternatives and relate them to their values and needs. One has to reconsider that multiplication of expert opinions shows no simple mistakes in using the correct method, but an "alternative" approaching to the same problem at stake. Expert opinions should be multiplied, in principle, as far as possible. "Lay" people, unable to participate in strategic knowledge production of these fields themselves, except via the mentioned roles of complementing or falsifying expert knowledge in trivial ways, get an overview of comprehensive issues through expert debates. The technique of bringing them together with a group of experts providing if possible, different expertise on the topic at issue and making the ideas understandable for the citizens may bridge over the difficulties when only scientific instrumentation is appropriate for providing detailed knowledge.

Let me come back, after acknowledging that the "lay" citizens may simply falsify some knowledge of experts to ask whether it should be accepted that they really are "experts of their own life". So far the comparison is between the limits where simple falsifying actions and observations can be made and "experts" simply put models on their life situation from outside, the roles can be trivially questioned. At least "experts" may be proven more or less lacking the knowledge of decisive local conditions. This does not mean, of course, that residents, users of some technology are "experts of their own situation", but they may know something still unknown by the experts. Dialogue research, interactive social science research are approaches to overcome this. But committing to modernization of life, where stabilised frames of life are regularly "attacked" by new technologies the "lay" citizens certainly need the interaction with appropriate advisorship.[40] (It is typical, I think, that the most different types of experts appear in the everyday lives of those who are able to pay for it.) Even when it does not exceed the interaction with market researchers the interaction of advisers and the "lay" may make the demarcation to become rather dim. Experiences gathered in the service industry show the dynamic nature of managing the problems of developing and disseminating new services and show how much a need for mutual learning by developers and "lay" users is evident.

It is important to recognize that often citizens are ready to learn. This readiness can include getting acquainted with strategic knowledge, and also learning about their own values. Dialogue research provides for a case for this non-focal effect of the interaction process with experts as learning of "lay" people about their own values. The famous risk researcher Keeney recognized that the use of the "public value forum", a technique to illuminate and clarify public values leads to some non-focal result, too, that can be developed into an additional purpose for the investigations. "Lay" people began spontaneously to learn from expert criticism about inconsistency and neglected fields of their knowledge.

Taken from another point of view people are innovation resources in a high-tech society so far as the interface with high-tech instruments will be necessarily reproduced. "Learning by doing" remains largely for "lay" people. Users are not givens but co-produced. Their local investigation capacity to check and to set right technologies at the locally different places systematically produces knowledge about the localities, and some of this knowledge may initiate new generalisations, recognition of essential features of technologies. Learning by doing is the typical form of innovation in these cases and the ordinary customers are the resource of new knowledge. But not only are users not ready made, expert knowledge too is a result of a piecemal generalization process. A spiral like dynamics of the interaction of scientific (science, social science, humanities) experts and the "lay" citizens displays itself. It is a dynamics of modelling and contextualization, in which at every state a mutual criticism may move the dynamics toward higher levels. Production of hybridspaces as loci of interaction are the typical nodes.

As mentioned already it would have been easy to finish here, some years ago. From then there is a recognition vindicated of overall importance. It refers to the needed reconsideration of judgements on the qualitative characteristics of risky situations.

Let me summarize the claim of the already several times mentioned PESTO paper. The authors reconsider the relation of risk science, precaution and participation by the citizens. Their starting point is that any knowledge of complex issues is a somewhat arbitrary reproduction, modelling of the is-

---

[40] One can consider the different advisers visiting him/her with marketing purposes and may wonder about the possibilities of developing advisorship for truly emancipatory purposes.

sue. Risks are richer than having only quantitative characteristics, expressed by the relation between magnitude and probability. They are also severe, or/and immediate, or/and reversible, or/and differently distributed in space, different from the point of view of intergenerational equity, controllability etc. These should be taken as a sort of "hidden variables". Different cultural groups, political constituencies or economic interests will typically attach different importance to the different forms and dimensions of risk. "The apparent riskiness of different options might be expected to vary quite radically, depending on the priorities attached to the 'hidden variables' during the process of appraisal."[41] It is not to say that "anything goes" and can have a rightful truth claim. But many approaches are of the same right, according to the chosen parameters. The chosen approaches are appropriate representations for some goals, inappropriate for some others. The authors claim that participation gets with this a quite new meaning. One can not only look for its essential place because social action is necessarily "underdeterminated" by expert knowledge but cognition itself is also "underdeterminated" without the needed compass, the goals, goal combinations that determine what characteristics of risky situations should be taken into account. They conclude that the citizens, their groups, I would say all the relevant actors in the risk arena, must have their necessary place – already in framing cognition.

As mentioned the authors of the IPTS material seem to have made this conclusion. Actually, I think that their conclusion is a short cut. The problem is the following. They argued convincingly that without a choice of values to frame the risk research process it cannot work. But it depends on the nature of the political relations, exemplarily modelled by the three idealtypes of Habermas, that, to solve the problem, in one case would mean a simple step toward strengthening expertise by involving experts of values. The "libertarian" would offer a mode of policymaking in which leaving the most possible task to the experts would arguably bring the most epistemic profit, while a "communitarian" mode would be a model for realization of the conclusion made by the authors of the IPTS material. Because value commitments are unavoidable on the evaluatory level, too, defenders of the one position would emphasize the efficiency of their way, defenders of the other position would emphasize qualitative criteria. Let us take then the model they offer. This model would allow for a co-operation for experts and the "lay" citizens to make definite the risk issue at stake. So, as they claim, including the citizens in the decisions on the framing assumptions would be a consequence of making open deliberations on possible framing value choices, so raising the scientific character of the investigations. I oppose considering this that the authors have not been careful enough. With this I also want to say that it seems there is no simple epistemic necessity in itself to involve the "lay" citizens in the cognitive process, but, preserving an essential part of their argumentation there may be a good mixed, political and epistemic reason

---

[41] Stirling (1999) p 21.

to do this way. In a type of "deliberative" democracy a dialogue realized by "value experts" and the citizens may warrant the most careful inclusion of the citizens.

The problem is potentiated in issues of risk research by a further one. It is in order to mention another problem, elegantly discussed by Funtowicy and Ravetz (1990). To summarize: the problem arises from comparing risky situations through the decisions at stake and the magnitude of uncertainty. It comes out that in one field small decision stakes and small uncertainties are identified, while another field also exists, where the growing decision stakes are paired with growing uncertainty. Here, it seems the co-operation of citizens and experts is becoming rather essential, experts function in an advisory role.[42] There is then a third field, too, where the stakes are enormous and the uncertainty is also rapidly growing. Funtowicz and Ravetz conclude that only citizens will be left here as actors. My objection is that, taking into account that the demarcation between the second and the third field in their representation of the issues determined by uncertainty and stake is moving, important place for experts would be recommendable in their model. This is their role of "housebreaking" situations in which uncertainty is not manipulable scientifically in an appropriate manner. If this "housebreaking" succeed, the dialogue between experts and citizens would regain place, and risk experts would regain some advisory role.

Risky situations are much richer and more multidimensional than it seems from the extremely reductive approach to them by quantitative risk assessment only. One layer was begun to be investigated by Funtowicz and Ravetz, another by the authors of the IPTS paper. It would be interesting to investigate the relation between the two cases and gain a clearer knowledge this way.

I think it is most important to recognise that many good arguments can be developed against pure forms of cognition. Technology Assessment is an especially good example as practical goal research in exemplifying this.

I think that the recognition of the many faceted role possibilities of the "lay" in TA, in technology policy and in constructing and using technologies are one side of a situation where to reconstitute trust in (a changed type of) expertise is simultaneously in order. Experts and "lay" persons are co-producing each other. Expertise is the most important, if it contains built in self-reflectivity as a readiness to recognize the limits, set by its decontextualizing nature that reproduces the need for "local knowledge". Recognition of the limits set by the incertitude of situations is essential for handling risky situations more flexibly, nevertheless while aiming at "robust" knowledge as far as it is possible and needed.

---

[42] They also speak about a third field where there is no possibility to have expert knowledge. This field is uninteresting for us according to the purpose of the recent paper.

The story of institutionalized Technology Assessment began and evolved through an expertocratic model, through a special type of dogmatism. The fight for citizen "participation" aimed at building opposition, as an immediate form of self-defence. Captured by the immediateness of their situation the actors in the policy arena behaved as if defined by contradictory interests only. The regulation of technological development has been based on output regulation, not on co-production and on passive knowledge of the effects of givens, production processes or products. Scientific knowledge of the situations, perceived in a reifying way, and lay perception, captured by the immediacy of experience tighted against each other. TA has recently moved toward deliberative forms as cTA, the constructivistic form of Technology Assessment. This new period emerged from the understanding that technologies can be better developed and can gain better public acceptance if the citizens (users, customers) are involved in the construction work. We are living now in something that could be described as a "third industrial revolution". One of its decisive pecularities, in my wish it should be the decisive feature, is that of reconciling innovation with social acceptability. It seems arguable that this goal can only be approached if a stabile co-operative learning and co-production process can be developed. One of the preconditions of this process has been some reflective learning on the cognitive processes in society. To these pieces of recognition belongs the knowledge of the distributed nature of knowledge, the self-reproducing necessary cognitive limits of expertise, and the intrinsically mixed nature of cognition and policing.

References

Collingridge D (1980) The Social Control of Technology. Open University Press, M. Keynes

Cronberg T et al. (eds) (1991)   Danish Experiment – Social Constructions of Technology. New Social Science Monographs, Institute of Organisation and Industrial Sociology Durant, Copenhagen

Fischer F (1990) Technology and the Politics of Expertise. Sage, Newburz Park, London, New Delhi

Frankenfeld PhJ (1992) A Normative Framework for Risk Studies. In: Science, Technology and Human Values. V17, N4, Autumn

Habermas J (1996) Die Einbeziehung des Anderen. Suhrkamp, Frankfurt/Main

Keeney et al. (1993) Decisions with Multiple Objectives. Wiley, New York

Krohn W (1997) Die Innovationschancen partizipatorischer Technikgestaltung. In: Köberle S, Gloede F, Hennen L (eds) Diskursive Verstaendigung? Nomos, Baden-Baden

McMullin E (1983) Values in Science. Philosophy of Science Association 1982, Asquith PD (ed), East Lansing, Phil. Of Sci. Association

Nennen HU (1998) Das Expertendilemma: Ein Fazit. In: TA-Informationen, 3/98

Perrow C (1984) Normal Accidents. Basic Books, New York

Sclove RE (1995) Democracy and Technology. The Guilford Press, NY

Shrader-Frechette KS (1991) Risk and Rationality. University of California Press, Berkeley, Los Angeles, Oxford

Shrader-Frechette KS (1993) Burrying Uncertainty. Univ. of California Press, Berkeley, Los Angeles, London

Stirling A (1999) ESTO Project Report On Science and Precaution in the Management of Technological Risk Raise. Sevilla

Thurow L (1999) Brainpower and the future of capitalism. In: Ruggles R, Holthouse D (eds) The Knowledge Advantage, Oxford, Capstone

Weinberg A (1972) "Science and Trans-Science". Minerva, 10(2), April 1972

Wynne B (1992) Uncertainty and Environmental Learning: reconceiving science and policy in the preventive paradigm. Global Environmental Change, 6/92

9 The Role of Scientific Input and Public Participation for Technology Assessment

Ortwin Renn

9.1
Introduction

Human beings depend on technology for survival and well being. Without organizational and technological intervention in nature, humankind would never have been able to maintain the population densities encountered today or sustain an ethical claim to an individual livelihood in dignity. At the same time, however, technology creates risks and negative side effects. The extent to which risks are associated with technological development and the judgment on the balance between risks and benefits are key elements of Technological Assessment (TA). It is the task of Technology Assessment to highlight the potential for positive and negative developments and suggest possible modifications or policy options with the aim of assisting decision makers to limit negative impacts to a tolerable level and to enhance positive opportunities.

In accordance with this task description, most policy makers expect TA experts to help design strategies that promise to prevent or mitigate negative and promote positive impacts of technological choices. In addition, TA expertise is demanded as an important input to design and facilitate communication among the different stakeholders in debates about technology and its impacts. Based on these expectations, scientific expertise in TA can assist policy makers to address five major functions (Renn 1995; Mohr 1996):

- Providing factual insights that help policy makers to identify and frame problems and to understand the situation *(enlightenment function)*;
- providing instrumental knowledge that allows policy makers to assess and evaluate the likely consequences of each policy option *(pragmatic or instrumental function)*;
- providing arguments, associations, and contextual knowledge that help policy makers to reflect on their situation and to improve and sharpen their judgment *(reflexive function)*;

- providing procedural knowledge that help policy makers to design and implement procedures for conflict resolution and rational decision making *(catalytic function)*;
- providing guidelines or designing policy options that assist decision makers in their effort to communicate with and to the various target audiences *(communicative function)*.

These five functions touch on crucial aspects of policymakers' needs: First, scientific insights help policymakers to understand the issues and the constraints of different policy options when designing and articulating policies. Policymakers need background information to develop standards, to ground economic or environmental policies on factual knowledge, and to provide information about the success or failure of policies. Second, scientific methods and their application are needed to construct instrumental knowledge in the format of "if – then" statements and empirically tested theories; this knowledge leads to the articulation of means-ends oriented policies and problem solving activities (Mohr 1994, p 197).

Thirdly, scientific reasoning and understanding help policy makers to reflect on their activities and to acknowledge social, cultural, institutional and psychological constraints as well as opportunities that are not easily grasped by common sense or instrumental reasoning. Fourth, policy makers may use scientists to design procedures of policy formulation and decision making in accordance with normative rules of reasoning and fair procedures. These procedures should not interfere with the preferences of those who are involved in the decision making process, but provide tools for making these preferences the guiding principle of policy selection. To meet this function, scientists need to play a role similar to a chemical catalyst by speeding up (or if necessary slowing down) a process of creating an agreement among those who are entitled to participate in the policy making process (Renn 1999). Lastly, scientific experts can help to design appropriate communication programs for meeting the purpose of legitimizing public policies as well as preparing target audiences for their specific function or role in the implementation of risk management tasks (Pidgeon 1997). This communicative function touches upon sensitive management privileges and is highly contested in the political arena.

The following article focuses predominantly on the delicate mix of scientific expertise and participatory instruments in the field of Technology Assessment. The second chapter will cover the different forms of legitimizing decisions in democratic societies and show how scientific input has a role to play in a concert with other forms of legitimizing collectively binding decisions. The third chapter will focus on the pitfalls and problems of using experts as consultants for policy makers. The fourth chapter specifies the requirements for a successful cooperation between experts and policy makers in the field of TA. The fifth chapter explains why scientific input has to be seen as one major element within a larger participatory framework of Technology

Assessment. The sixth and seventh chapter sheds some more light on the interaction between scientific input and participatory methods of Technology Assessment and evaluation. The last chapter summarizes the main results of this article.

9.2
Forms of Legitimization for Collective Actions

In a pluralistic society politics is faced with a dilemma with almost no way out where the question of legitimization is concerned. On the one side we find politicians with a mandate, as well as representatives of the authorities who prefer a certain solution, on the other side we find irreconcilable requirements and wishes of pluralist groups who, as a rule, reject the solutions proposed by politics and the administration. In turn many politicians and some administrations consider the counter-proposals of these groups problematic, illusory or not politically feasible. This situation is aggravated by the fact that varying opinions do exist within the public authorities and that social groups also do not represent a unified front, but are in turn split into manifold groups with various interests and values. New political alliances and pacts of convenience are repeatedly called to life, frequently motivated only by common interests. Moreover, conflicts which are difficult to solve flare up between those interested in the concrete solution of a pressing task at hand and those wanting to express, with each action, an abstract orientation for a future design of society (Rittel 1992, p 29).

Faced with this intricacy of demands, concepts and counter-concepts, political decision-makers have a hard time making appropriate and politically balanced decisions. Even if politicians struggle to make a decision in favor of one of the possible options, they inevitably get caught up in the crossfire of criticism due to the plurality of opinions and assessments. Depending on the type of decision, one or another group will accuse them of lacking expertise, bound interests, blindness or even cynicism – regardless for which of the possible variants they decided. In this dilemma, many politicians are inclined to delay making a decision until they are forced to act by external powers or they pass the buck of unpopular decisions to other institutions. Particularly popular is the passing on to advisory committees, which are either recruited from science or from the so-called relevant social groups (model of corporatism).

Political sociologists discuss the problem of "political legitimization as a scarce and ever scarcer resource" under the heading of "Steuerungsproblematik" (the problem of legitimization) (cf. Scharpf 1991; Mayntz 1993, pp 41–43; Willke 1995). Some analysts claim that modern societies such as ours can no longer practically be governed by political bodies. The overall system of society, they say, increasingly disintegrates into more or less autonomous subsystems (cf. Luhmann 1984). In this situation the only thing left to do is to make sure that there is mutual accessibility between the subsystems, i. e. a communication flow between politics and other subsystems (such as economy, science, community). An agreement on common goals, values or vision is highly unlikely due to the fragmentation of society into pluralistic subsystems. Opposed to that, other social scientists continue to rely upon the integrative power of system-spanning argumentation and rationality (cf. Habermas 1992). From this point of view, political legitimization depends on whether the prerequisites for a fair exchange of arguments (in the sense of a discourse) are fulfilled.

This more theoretical discussion however distracts from the fact that, at any point in time, decisions which regulate society are made continually. Some of these governing processes may be uncoordinated, they may happen rather coincidentally or they may be inconsistent in their results, but day after day far-reaching technological choices are made. Waiting out decisions or not wanting to commit oneself is also a decision, which has consequences.

The main technological choices are performed in abstract space by the interplay of the four central subsystems of a society with their special system logic (cf. Renn/Webler 1998, p 9ff). These four subsystems are the economy, science (expertise), politics (including the legal systems) and the social system. These four systems have several governance processes and structures adapted to the system properties and functions in question. The systems and their most important structural characteristics are shown as a scheme in Fig. 9.1 (cf. Habermas 1968; 1978, Lohmar 1967, pp 98ff; Kweit and Kweit 1981, Scarcinelli 1990).

What findings can be inferred from the comparison of these four systems? Economical steering mechanisms largely determine the actions in the economy, scientific methods the finding of new knowledge in the sciences, etc.. Included are also special mechanisms of governance in each sector:

1. In the market system, decisions are based on the cost-benefit balance established on the basis of individual preferences and willingness to pay.
2. In politics decisions are made on the basis of procedural methods of decision-making and norm control (within the framework of a given system of government).
3. Science has at its disposal methodological rules for distinguishing true statements from false ones, with the help of which one can assess decision options according to their likely consequences and side-effects.
4. Finally, in the social system, there is a communicative exchange of interests and arguments, which helps the actors to come to a jointly agreed solution.

With respect to the question of governance of modern societies, those mechanisms and instruments are significant which were developed within these four systems for the creation and justification of collectively binding decisions. Although collectively binding decisions are made mainly in the sphere of politics, the remaining three systems participate directly and indi-

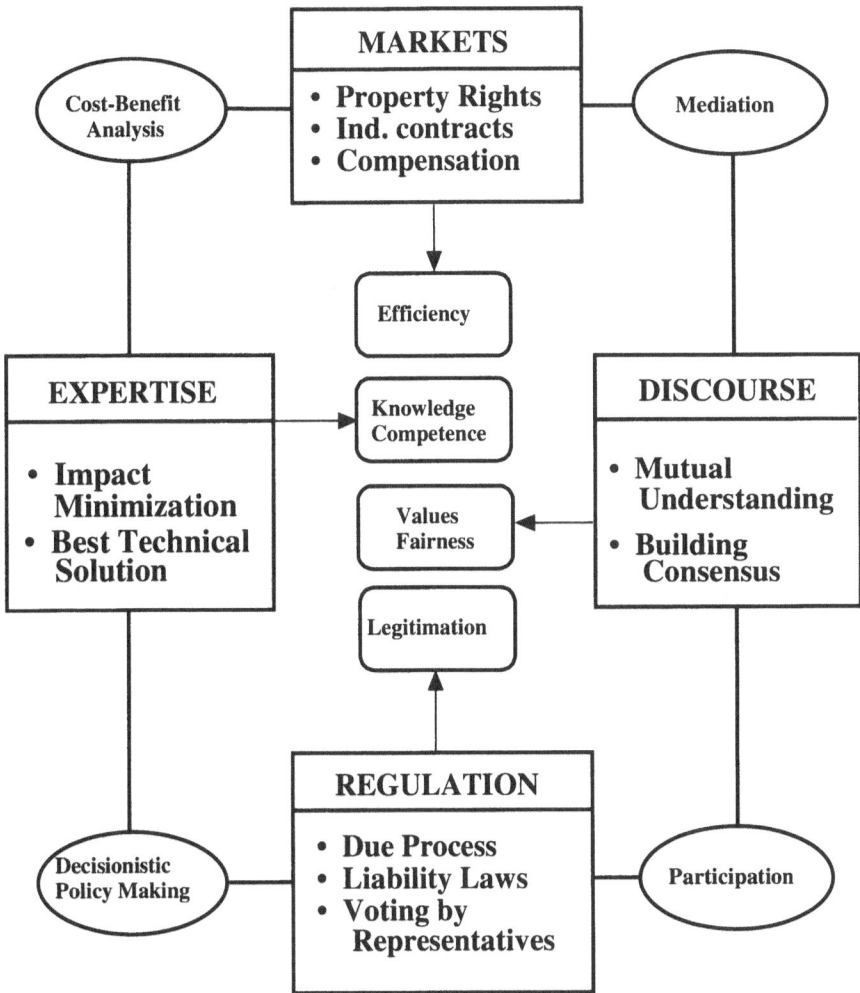

Fig. 9.1. The four central subsystems of a society

rectly, with their own instruments and mechanisms, in the creation of collective decisions.

Corresponding to their respective system logic, the four systems considered can be allocated certain methods or instruments, which are used basically or in hybrid forms when accomplishing political steering processes. Economic system logic has the instruments of (shadow) price setting, financial incentive systems, the transfer of rights of ownership of public or non-rival goods and financial compensation (damages, insurance) to persons, whose utilities have been reduced by the activities of others. The expert system makes use of a wide variety of knowledge-based decision methods (e. g.

choice of appropriate methodology, Delphi, meta-analysis, medical consensus conferences and others) in order to determine the likely and unlikely impacts of different decision options. In the social field, discursive forms of communication- and consensus-based decision finding are used within the framework of advisory committees or citizen participation procedures for political purposes. The conventional instruments of the political system comprise the jurisdictive processes of passing a resolution and of scrutinizing a resolution. Votes in a parliament are just as much part of this governance model as is the verification of decisions before a court.

Rarely, socially relevant problems are dealt with within the limits of one single system logic. Rather they go through hybrid procedures, which are to be seen as combinations of the individual systems. The settlement of conflict with the method of mediation or negotiated rule making can for example be interpreted as linkages between the market model and the social sector, while the cooperation between experts and political representatives in joint advisory committees (i. e. the experts provide background knowledge, politicians the preferred decisions based on it) represents a combination of knowledge-oriented elements and political governance. Classical hearings are combinations of expert knowledge, political resolutions and the inclusion of citizens in this process (for comparisons with these models see also the three political advisory models in Habermas 1968, and the steering models of power, money and knowledge in Willke 1995, and the observations of Renn/Webler, 1998 p 14ff).

9.3
Pitfalls and Illusions in the Cooperation between Experts and Policy Makers

What are the potential contributions of technical expertise to the policy process when it comes to technological choices or regulations? In principle, experts can provide knowledge that can help to meet the five above-mentioned functions and to anticipate potential risks before they materialize. But they can do this only to the degree that the state of the art in the respective field of knowledge can provide reliable information that pertains to the policy options. Many policy makers share certain assumptions about expertise that turn out to be wishful thinking or illusions (Boehmer-Christiansen 1997). Most prominent among them are:

- *illusion of certainty*: making policy makers more confident about knowing the future than is justified (Funtowicz and Ravetz 1990, pp 17ff.);
- *illusion of pseudo-certainty*: making policy makers overconfident that certainty in one aspect of the problem applies to all other aspects as well (Kahneman and Tversky 1979);
- *illusion of "absolute" truth*: making policy makers overconfident with respect to the truthfulness of evidence (Ravetz 1989);

- *illusion of ubiquitous applicability*: making policy makers overconfident in generalizing results from one context to another context (Rip 1985; 1992).

The experts themselves often reinforce these illusions. Many experts feel honored to be asked by powerful agents of society for advice. Acting under the expectation of providing unbiased, comprehensive, and unambiguous advice they often fall prey to the temptation to oversell their expertise and provide recommendations far beyond their realm of knowledge. This overconfidence in one's own expertise gains further momentum if policy maker and advisor share similar values or political orientations. As a result policy makers and consultants are prone to cultivate these illusions and act upon them.

The second major problem in using scientific expertise for policy making lies in the difference between the two systems with respect to values and aspirations (cf. Habermas 1978, pp 120ff.). Science is knowledge-oriented, policy making is action-oriented. As trivial as this may appear on first glance, it has several implications with respect to value conflicts. Among them are (Rip 1985, pp 94ff. or Mayntz and Scharpf 1973, pp 115ff.):

- *administrative efficiency versus public scrutiny*: Scientific research is based on open discussions. Without a debate among the research community, claims for truth cannot be sustained. Policy makers often fear a public debate because it delays the process of decision making and provides information about potential alternatives to economic or political competitors.
- *Claims for truth versus political interests*: Regardless of the fact which school of science theory one adheres to, scientific statements are meant to represent claims of truth. These claims may be clouded by personal interests and values held by the researchers or their respective institutions, but in a situation of value conflicts between personal interests and (subjectively) recognized truth, the norm for scientists is to give "truth" priority. Policy makers, on the other hand, are obliged to act in the interest of their clients and to search for arguments that justify the pursuit of this interest. Truth and interest share some common ground (a wrong prediction may also hurt those in which interests the policy maker acts), but they often are in conflict with each other (cf. case study in Knoepfel and Weidner 1982).
- *Academic rigor versus time constraints for needed actions*: Policy makers act under time constraints. Doing nothing does not solve the problem; it may even aggravate it. Scientists need time to explore the various options, reduce uncertainty, and find reliable answers to the questions of the policy makers. In almost all cases it takes more time to do thorough research than it is possible to delay the decision (Lynn 1986).

For both reasons, the inherent limits of using systematic knowledge for policy processes and the value conflicts between science and politics, the cooperation between experts and policy makers is necessarily linked to potential conflicts and misinterpretations. Any economic or political system needs to deal with these problem of incongruity and develop structures and procedures that help to take advantage of the potential input of science and at the same time respect the inherent limitations as well as the differences in values and orientations. This issue will be further developed in the next chapter looking specifically in the role of expertise for Technology Assessment.

9.4
The Challenge of Integrating Expertise into Technological Policy-Making

9.4.1
The ambivalent role of science for legitimizing decisions on TA

Several sociologists of science and knowledge have analyzed TA-decisions as they are transformed from professional policy issues to topics of public debate. For example, Austrian sociologist Helga Nowotny and some of her colleagues have studied the role and function of scientific information for risk management (Nowotny 1979; Nowotny and Eisikovic 1990; Evers and Nowotny 1987). They pointed out that the economic and political systems demand reduced uncertainty. Thus, both systems place severe demands upon the scientific community to make uncertain aspects of economic and political actions less threatening. According to Nowotny and her colleagues, there are two major problems connected with the quest for more certainty. First, the two main characteristics of modern science, specialization and abstraction, tend to promote overconfidence in the results of scientific assessments. Specialization has lead to relevant factors being excluded from the analysis; abstraction to systematically underrating idiosyncratic events. Second, the results of scientific risk assessments are used to support certainty even if scientific evidence is not conclusive.

German sociologist Peter Weingart has followed up on this thought and linked the Technology Assessments to the need of society to constrain the realm of potential options as a means to manage its complexity (Weingart 1979; 1983). He assumes that scientists have lost a monopoly on providing systematic knowledge to society. He asserts that, as science looses its privilege to speak the "ultimate" truth, it can better serve its latent integrative function. As Weingart points out, however, the use of scientific methods and arguments is not arbitrary. The internal rules of science may be yet too weak to establish or refute the validity of plural scientific claims. They are, however, Weingart argues, strong enough to exclude options and actions that are in contradiction with basic principles established by a broad consensus on valid scientific methods and what they can produce. Both the science system and the political system benefit from this reduction potential.

A less optimistic view on the role of scientific arguments in TA can be found in the work by American sociologist Sheila Jasanoff (1986; 1993). Although she agrees with the arguments by Weingart and others that all groups in the risk arena use risk analyses to their own advantage and rely to a large degree on scientific or pseudo-scientific arguments, she notices, however, a rather selective use of these arguments by the decision making bodies. If, as she believes, science is producing constructions of reality, it does indeed provide arguments to all possible claims. But this also means that the decision-making bodies have the liberty to select one interpretation over the other. In a recent article (1998), she distinguishes three perspectives: a realist, constructivist, and discursive view on the use of scientific assessments in policy arenas. Each of the three perspectives leads to another set of selection rules for accepting or refuting expertise. Jasonoff's analysis is backed up by the work of Tom Dietz and others, who investigated the role of experts (whom they called risk professionals) in risk debates (Dietz and Rycroft 1987; Dietz et al. 1989). They found a devoted group of pro-governmental scientists who basically are eager to find supporting evidence for predefined political positions. At the same time, so the analysis by Dietz and Rycroft, opposing experts have developed their own professional network and research agenda, which, as it grows, become more capable of launching counter-attacks and producing more public pressure onto the political elite.

American political scientist Dorothy Nelkin has come to the conclusion that science has become a supermarket for rationalizing political decisions (Nelkin 1977; Nelkin and Pollak 1980). The plurality of scientific expertise jeopardizes both the status of scientists and the credibility of politicians. Each of these legitimization losses can paralyze the political system and lead to a situation of indecisiveness and the preference for postponing painful decisions. Nelkin believes that the political system turns to increased participation under these circumstances not because it believes in the wisdom of the public. It does so because it realizes that it must make its choice of actions transparent to the major actors in society and gain public support for their actions.

9.4.2
Lessons for the function of science for Technology Assessment

What insights can one gain from this brief review of some prominent sociological analyses with respect to the role and function of science in the field of TA? The need for decision makers to make a positive choice among different options of technological actions conflicts obviously with the claim of most interest groups to legitimize their specific viewpoint through scientific reasoning. To choose among equally legitimate courses of action becomes an almost insurmountable task since no meta-arguments are available or convincing enough to distinguish valid from invalid claims (Renn 1995). This is particularly true for debates on technologies since the prognostic nature of reasoning prohibits the application of simple rules for falsifying claims. Even if a nuclear reactor will experience a serious meltdown within the next five years this does not mean that the usual risk assessments calculating an average probability of 10^{-3} for the occurrence of a meltdown are obsolete or wrong. In this situation of uncertainty and ambiguity, resolution of scientific debates can proceed in five different ways:

- First, the choice among competing truth claims is made solely on political grounds based on the persuasive power of the decision making body (such as a proposer or an agency). Over the last two decades this reliance on power and persuasion has not been successful and lead policy makers to increased legitimization problems (Jasonoff 1982). The reason is that in secular and value-pluralistic societies there are only few normative principles that are accepted by all relevant actors (Kemp 1985, Renn 1995). In spite of these legitimization problems, reliance on political power and selective use of expertise are still dominant means of resolving risk conflicts in many Southern European societies (O'Riordan and Wynne 1987).
- Second, the selection is based on formal criteria for selecting expertise or more frequently experts. Such formal criteria can be constructed according to strategic goals such as finding arguments for decisions already taken or preconceived. Most often policy makers use status or prestige for justifying their selection of experts (Beck 1992). This system works both ways since the experts who are close to the decision-maker's point of view may also be most eligible to receive the necessary credentials. It is obvious that such a strategy may not produce results that are most beneficial to society. In addition, groups dissatisfied with the selection process will demand that "their" experts are included in the experts' club. Either the political decision-makers are powerful enough to resist such efforts (which would reaffirm solution 1) or they open the decision-making arena to conflicting interests and science camps (Coppock 1985).
- Third, the selection is transferred from the political into the legal arena. This strategy has been very popular in the United States but also in Europe and lately in Japan. Although the legal system normally succeeds in producing unambiguous results, there is a growing number of critical reviews emphasizing a lack of competence and fairness associated with legal processing of scientific controversies (Majone 1989; von Schomberg 1992). The typical legal process of cross-examination benefits those parties that can make an easily accessible and communicable knowledge claim, are professionally prepared for legal argumentation, and have the financial and social resources to provide legally acceptable evidence. In addition, legal processes are both costly and time-consuming (Folberg and Taylor 1984).
- Fourth, the selection is left to a "muddling through" process (Lindbloom 1959). All parties should be able to play out their muscles and to push their point of view into the public arena. The rationale behind this strategy is to trust the "invisible wisdom" of the pluralistic information and communication market and to expect an optimal outcome based on free competition among rival claims (Olson 1984). This strategy has several problems: it as-

sumes equal or proportional power to all parties, it takes for granted that the rationality of the economic market is equally valid and functional in the political sphere and it reduces questions of truth to bargaining power.

- Finally, the selection can be made in a mutual communicative process in which the participants agree in advance to follow specific rules of validating or falsifying competing knowledge claims and to argue over different equally valid strategies to resolve uncertainties and ambiguities (Amy 1987; Susskind and Cruishank 1987; Dryczek 1990; Webler 1995). Such a communication process based on analytic-deliberative reasoning has been recommended by many observers of the TA community and has found enthusiastic support by many policy-making bodies (Fiorino 1989; President's Council 1997; Harter et al. 1998). There is, however, no model or formal procedure available that has proven to deliver what the proponents of analytic-deliberative processes expect from such models. New procedures are necessary to provide a structure for a communication and deliberation process that facilitates decisions on technological choices (or any other issues involving scientific input) that meet the criteria of competence, fairness, legitimacy, and cost-effectiveness (Renn et al. 1993; Renn and Webler 1998). So far many models have been suggested and a few have been tested to a limited degree. Most societies are still far away from a comprehensive evaluation of the existing models, let alone a conclusive inference on what model to use for what purpose.

9.5
The Requirements for a Model Linking Expertise to Technology Assessment

What are the requirements that would satisfy the need for a competent, knowledge-based, fair and deliberative process of assessing and evaluating technological impacts? Based on the analyses from theorists of human knowledge and science that had been summarized in the previous sections one can draw the following inferences on the required process characteristics that need to be met when making complex technological choices:

- Regardless whether one prefers a constructivist or realist perspective on human knowledge about technological impacts (cf. Bradbury 1989; Horlick-Jones 1998), scientific rationality as defined by methodological consensus among risk researchers is insufficient in making unambiguous and uncontested claims about the characteristics and severity of the specific technology under investigation (Margolis 1996).
- In analyzing technological potentials, one needs to include systematic and anecdotal sources of knowledge (Wynne 1989). Systematic knowledge is necessary to build upon the collected experiences of the past, anecdotal knowledge to take account of the idiosyncratic features surrounding the specific decision problem.

- When contemplating the impacts and consequences of developing or applying technologies, one needs to be informed about the likely consequences
of each decision option and to be cognizant of the potential violations of
interests and values connected with each decision option (McDaniels 1996).
Although both steps, predicting the likely impacts and evaluating the de-
sirability of each of these consequences, can be separated analytically it
is counterproductive to run the two processes in parallel and assign these
tasks to different agents, since the answers of the first task co-determines
the answers to the second task and vice versa (Primack and von Hippel
1974; Renn 1995). What is needed is a procedure that integrates both
tasks without sacrificing the necessary precision and quality of factual and
normative judgments that are inherent in both steps.
- Integrating values into technology evaluation requires the input of those
people whose interests and values are affected by the decision options (Pidgeon 1997; Kunreuther and Slovic 1996). In many instances, these interests
and values are so obvious that agencies can act on their behalf without major reassurance that their action is in accordance with the needs and concerns of those whom they serve (Chess et al. 1998). In many technological
decisions, however, it is less obvious what is in the best interest of the people and plural value input is needed to produce a fair and balanced decision
(Creighton et al. 1998). If only divergent interests need to be reconciled, involvement of stakeholders may suffice; if broad value judgments or issues of
social justice are addressed, representatives of the affected public ought to
be involved. In both cases such input requires direct participation efforts
beyond the scope of normal decision making procedures based either on
agency rules or majority votes by a representational branch of government
(Chess et al. 1998; Webler and Renn 1998).

To identify public values and integrate facts and values into a joint decision making effort, a communication process is needed that build upon intensive dialogue and mutual social learning. Without consulting public interest
groups and those who are affected by the decision, a meaningful synthesis of
expertise and public concerns cannot be accomplished. A recent report by
the National Academy of Sciences calls for an integration of assessment and
discourse forming an "analytic-deliberative" approach (Stern and Fineberg
1996). Using similar arguments, modern theorists of public planning have
questioned the validity of the traditional paradigm of instrumental rationality, which assumes that experts and public servants have almost perfect
knowledge about the most likely outcomes of different planning options as
well as about public preferences with respect to each bundle of outcomes
(Forester 1989; Sager 1994). The new paradigm of transactive planning or
disjointed incrementalism is based on procedural rationality, which builds
on communicative actions. The objective is to design cooperative planning
processes in which uncertain outcomes are discussed with representatives of
the affected public and the evaluation of options is performed in an active

dialogue between experts, stakeholders, and members of the general public (Fiorino 1990; Hadden 1989).

9.6
The Necessity of Discursive Processes when Dealing with Technological Impacts

Technology Assessment relies on a discursive process of acquiring and evaluating tested knowledge (cf. Evers and Nowotny 1987, pp 244 ff.; Baron 1995). The fact that a subject is intensely talked about does not make it a discourse. Discourses are symbolic or real places of communication, where speech acts are examined for their claims of conclusiveness in a mutual exchange of arguments, according to predetermined rules of validity, without regard to person or status (Habermas 1971; 1981; 1992). The claims of conclusiveness presented in the discourse do not only refer to cognitive statements, but also comprise expressive (affects and promises) as well as normative observations. Ultimately, the discourse with its highly diverse acts of speech should reflect the diversity of the world of experience and its limitations (Böhler 1995).

Discourses are no panacea for all the problems of our times. They also cannot solve the problems of uncertainty and ambiguity in the research of technological impacts and assessment (Giegel 1992). The fact that parties in conflict meet at a round table and talk has, by itself, hardly contributed to clarifying facts, reaching new insights or solving conflicts. Rather it is essential that in a discursive procedure factual issues are clarified on the basis of a coherent methodology, that questions of assessment are discussed and action consequences are derived consistently (Renn and Webler 1995).

Discourse and an orientation of consent are frequently misunderstood by the public. "Just another blather joint", some state, "another proof of the lack of leadership in politics", others say (Weinrich 1972). Seen against the practice of many discourses, both reproaches are justified, but they miss the internal logic and immanent efficiency of discursive processes. Discourse does not mean agreement on the smallest, mostly trivial denominator. Rather, the point is a creative process where arguments are exchanged in all clarity and, if need be, sharpness and the different values and interests are presented. Frequently these discourses do not conclude with a consensus but with an agreement to disagree. In this case all participants know, why one side is for one measure and the other is against it. The individual arguments were then checked within the discussion and sounded out for weaknesses and strengths. The remaining differences are no longer based on specious conflicts or false judgments, but on clearly defined differences in the assessment of decision effects (Schimank 1992). The result of a discourse is more clarity, not necessarily unity.

9.7
Classification of Discourses

In literature many classification systems for discourses are to be found (Burn and Überhorst 1988; Zilleßen 1993; Rowe and Frewer 2000). One can argue about factual issues, assessments, action requirements or about aesthetic concepts. For practical work in Technology Assessment a classification into three discourse categories seems helpful to me (Wachlin and Renn 1999):

The *epistemological discourse* comprises communication processes, where experts of knowledge (not necessarily scientists) grapple with the clarification of a factual issue. The goal of such a discourse is the representation and explanation of a phenomenon as close to reality as possible. The more complex, interdisciplinary and uncertain this phenomenon is, the more a communicative exchange among experts is required to reach a coherent description and explanation of the phenomenon. Frequently, these discourses can only show the range or bandwidth of the still methodically justifiable knowledge, i. e. stake the claim where dissent can still be founded by methodic rules and/or empirical evidence.

The *reflexive discourse* comprises communication processes dealing with the interpretation of factual issues, the clarification of preferences and values and the normative assessment of problematic situations and proposals. There discourses are mainly suitable as opinion barometers for technical developments, as auxiliaries for a decision preparation and as an instrument for anticipative conflict avoidance. They convey an impression of moods, wishes and uneasiness, but without individually assessing concrete decision options.

The *creative or designing discourse* comprises communication processes aiming at the assessment and evaluation of action options and/or the solution of concrete problems. Procedures of mediation or direct citizen participation can be classified in this category as well as future workshops, consensus conferences and political or economic advisory groups, who are supposed to propose or evaluate concrete political options.

All three forms of discourse form the fabric of political consultancy in the narrower sense, for the results of discourses must be tied into legitimate forms of decision finding. Even if these discursive procedures succeed in providing recommendations in a result-oriented and efficient way, they will still not produce acceptable solutions, if the problems of complexity, ambiguity and uncertainty are themselves not made into subjects of discussion. Technology users and affected alike must be made quite clear, that risks are involved with each application of technology, and that harm cannot be excluded even with the best of intentions and greatest precaution. This may not be an excuse for incorrect behavior on the part of the institutions responsible for safety. But all people concerned must be quite aware of the fact, what is involved with a new technology and which potentials exist – good and bad. No warranties should be provided, at the most compensation in the sense of liability and insurance should be considered. Only by being fully aware of the remaining risks does the way open to new strategies of dealing creatively and precautionarily with ambiguity and uncertainty.

How to lead these discourses in the individual case is not the subject of this paper. In the past, we have gathered a lot of information about this (Renn and Webler 1998). In my home institution, the Akademie für Technikfolgenabschätzung (Center of Technology Assessment) in Baden-Württemberg, Germany, we consciously practice the path of discursive communication and try to lead all three types of discourse in parallel. We are still in an intense learning phase and some of these discourses have broken down prematurely. Nevertheless I am of the firm opinion that nothing but the discursive approach to Technology Assessment will show society a factually competent and morally justifiable path to the appropriate handling of technological assessment-options.

9.8
Concluding Remarks

The economic and political structures in modern societies undergo rapid transitions at present times. This transition is accompanied by increased pluralism of positions, values and claims, erosion of trust and confidence in governing bodies, increased public pressure for participation, and tendencies towards polarization between fundamentalist value groups and progressive change agents. The resulting conflicts can paralyze a political system if it proves incapable of integrating different outlooks and visions of the future and providing justifications of its decisions on the basis of both values and facts (Charnley 2000). At the same time, such conflicts can be catalysts of necessary changes that societies need to advance politically and economically. In this situation, Technology Assessment institutions are in urgent need to provide best scientific advice and assistance to policy makers.

Technology Assessment comprises both the scientific assessment of possible secondary effects and the evaluation of these effects according to the preferences of the persons affected, whereby both tasks, scientific exploration of potential effects and evaluation of these effects will remain blurred in their results due to the unavoidable complexity, ambiguity and uncertainty associated with each decision option. Prognoses concerning the technical future constitute a part of Technology Assessment and simultaneously are essential components of current decisions; they may not, however, pretend to promise that TA-experts can predict all dangerous events and developments and can thus also preclude them by preventive action. A more realistic picture of the potential scientific input to technological policy making is necessary in order to make Technology Assessment a vital and rewarding tool for policy processes. Among the major insights that I discussed in this paper have been:

- Scientific expertise can serve five functions: enlightenment, instrumental or pragmatic knowledge, reflection, catalytic process promotion and communicative exchange. All five products are badly needed for making prudent policy judgments and should be included in all models for using expertise in policy arenas. Such input, however, needs to be controlled by democratic institutions and should be open to public scrutiny.

• Scientific expertise is also used for legitimizing decisions and justifying policies that may face resistance or opposition. Expertise can therefore conflict with public preferences or interests. In addition, policy makers and experts pursue different goals and priorities. Expertise should be regarded as one crucial element of policy making among others. Scientific advice is badly needed, but its potential contributions should not be overrated. In particular, scientific expertise cannot and should not replace public input in the form of locally relevant knowledge and social values and preferences.

• Scientific expertise is absorbed and utilized by the various policy systems in different ways. It is essential that scientific expertise is integrated in a more "system and problem oriented" approach to policy making, in which science, politics, and economics are linked by strategic networks.

• Organizing and structuring discourses technological choices and policies goes beyond the good intention to have the public involved in risk decision making. The mere desire to initiate a two-way-communication process and the willingness to listen to public concerns are nor sufficient. Discursive processes need a structure that assures the integration of technical expertise, regulatory requirements, and public values. These different inputs should be combined in such a fashion that they contribute to the deliberation process the type of expertise and knowledge that can claim legitimacy within a rational decision making procedure (von Schomberg 1995; Charnley 2000; Renn and Webler 1998). It does not make sense to replace technical expertise with vague public perceptions nor is it justified to have the experts insert their own value judgments into what ought to be a democratic process.

What do these insights imply for the implementation of Technology Assessment? Firstly, Technology Assessment must differentiate functionally between scientifically identifying possible effects and their evaluation, while at the same time discursively interlinking both steps. TA should provide for a step-by-step, high-feedback and reflexive course of action in the weighing up of positive and negative effects by experts, users and affected citizens. Technology Assessment realized in such a way requires a close connection to the investigation of likely impacts and their evaluation without, however, giving up the functional and methodical differentiation between these two tasks (gaining knowledge and constructing preferences). As important as it is not to mix the methods of knowledge acquisition and impact evaluation, it is equally important to acknowledge the tight linkage between these two areas, as the research of technological impacts would otherwise no longer be efficient and realistic in an uncertain world. The necessity of this linkage also speaks for a discursive form of Technology Assessment.

It seems justified and desirable to initiate a common dialogue or discourse that bring together experts, policymakers, stakeholders and representatives of the affected public(s). However, a structure of discourse is necessary that makes sure that each group adds to the process the type of expertise, value judgments and reasoning for which they are the most qualified. This is particularly necessary if highly controversial subjects are at stake. The main lesson of past experiences with deliberative processes has been that scientific expertise, rational decision making, and public values can be reconciled if there is a serious attempt to integrate them.

Acknowledgement
I am deeply grateful for the professional and enthusiastic support by Ms. C. Rhodes for translating and editing parts of this article.

References

Amy DJ (1987) The Politics of Environmental Mediation. Cambridge University Press, Cambridge and New York

Baron W (1995) Technikfolgenabschätzung-Ansätze zur Institutionalisierung und Chancen der Partizipation. Westdeutscher Verlag, Opladen

Beck U (1992) Risk Society: Toward a New Modernity. Translated by Mark A. Ritter. Sage, London

Boehmer-Christiansen S (1997) Reflections on Scientific Advice & EC Transboundary Pollution Policy. Science and Public Policy 22, No. 3: 195–203

Böhler D (1995) Ethik für die Zukunft erfordert Institutionalisierung von Diskurs und Verantwortung. In: Jänicke M, Bolle HJ, Carius A (eds) Umwelt Global. Veränderungen, Probleme, Lösungsansätze. Springer, Berlin, pp 239–248

Bradbury JA (1989) The Policy Implications of Differing Concepts of Risk. Science, Technology, and Human Values 14, No. 4: 380–399

Burns TR, Überhorst R (1988) Creative Democracy: Systematic Conflict Resulution and Policymaking in a World of High Science and Technology. Praeger, New York

Charnley G (2000) Democratic Science: Enhancing the Role of Science in Stakeholder-Based Risk Management Decision-Making. Health Risk Strategies, Washington, D.C.

Chess C, Dietz Th, Shannon M (1998) Who Should Deliberate When? Human Ecology Review 5, No. 1: 45–48

Coppock R (1985) Interactions Between Scientists and Public Officials: A Comparison of the Use of Science in Regulatory Programs in the United States and West Germany. Policy Sciences, 18: 371–390

Creighton JL, Dunning CM, Delli Priscoli J (eds) (1998) Public Involvement and Dispute Resolution: A Reader on the Second Decade of Experience at the Institute of Water Resources. U.S. Army Corps of Engineers. Institute of Water Resources, Fort Belvoir

Dietz T, Rycroft RW (1987) The Risk Professionals. Russel Sage Foundation, New York

Dietz T, Stern PC, Rycroft RW (1989) Definitions of Conflict and the Legitimation of Resources: The Case of Environmental Risk. Sociological Forum 4: 47–69

Dryzek JS (1990) Discursive Democracy. Cambridge University Press, Cambridge, MA

Evers A, Nowotny H (1987) Über den Umgang mit Unsicherheit. Die Entdeckung der Gestaltbarkeit von Gesellschaft. Suhrkamp, Frankfurt am Main

Fiorino DJ (1989) Technical and Democratic Values in Risk Analysis. Risk Analysis 9, No. 3: 293–299

Fiorino DJ (1990) Citizen Participation and Environmental Risk: A Survey of Institutional Mechanisms. Science, Technology, and Human Values 15, No. 2: 226–243

Folberg J, Taylor A (1984) Mediation. A Comprehensive Guide to Resolving Conflicts Without Litigation. Jossey-Bass Publishers, San Francisco

Forester J (1989) Planning in the Face of Power. University of California Press, Berkeley 1989

Funtowicz SO, Ravetz JR (1990) Uncertainty and Quality in Science for Policy. Kluwer, Dordrecht and Boston

Giegel HJ (1992) Kommunikation und Konsens in modernen Gesellschaften. In: Giegel HJ (ed) Kommunikation und Konsens in modernen Gesellschaften. Suhrkamp, Frankfurt am Main, pp 7–17

Habermas J (1968) Technik und Wissenschaft als 'Ideologie'. Suhrkamp, Frankfurt am Main

Habermas J (1971) Vorbereitende Bemerkungen zu einer Theorie der kommunikativen Kompetenz. In: Habermas J, Luhmann N (eds) Theorie der Gesellschaft oder Sozialtechnologie. Was leistet die Systemforschung? Suhrkamp, Frankfurt am Main, pp 101–104

Habermas J (1978) Verwissenschaftlichte Politik und öffentliche Meinung. In: Habermas J (ed) Technik und Wissenschaft als 'Ideologie'. 9th Edition. Suhrkamp, Frankfurt am Main

Habermas J (1981) Theorie des kommunikativen Handelns. Bd. 1 & 2. Suhrkamp, Frankfurt am Main

Habermas J (1989) Erläuterungen zum Begriff des kommunikativen Handelns. In: Habermas J (ed) Vorstudien und Ergänzungen zur Theorie des kommunikativen Handelns. 3rd Edition Suhrkamp, Frankfurt am Main, pp 571–606

Habermas J (1992) Faktizität und Geltung. Beiträge zur Diskurstheorie des Rechts und des modernen Rechtsstaates. Suhrkamp, Frankfurt am Main

Hadden S (1989) A Citizen's Right-to-Know: Risk Communication and Public Policy. Westview Press, Boulder

Harter P, Orenstein S, Dalton D (1998) Better Decisions through Consultation and Collaboration. U.S. Environmental Protection Agency (EPA), Washington D.C.

Horlick-Jones T (1998) Meaning and Contextualization in Risk Assessment. Reliability Engineering and Systems Safety 59: 79–89

Jasonoff S (1982) Science and the Limits of Administrative Rule-Making: Lessons from the OSHA Cancer Policy. Osgoode Hall Law Journal, 20: 536–561

Jasonoff S (1986) Risk Management and Political Culture. Russell Sage Foundation, New York

Jasonoff S (1993) Bridging the Two Cultures of Risk Analysis. Risk Analysis 13, No. 2: 123–129

Jasonoff S (1998) The Political Science of Risk Perception. Reliability Engineering and Systems Safety 59: 91–99

Kahneman D, Tversky A (1979) Prospect Theory: An Analysis of Decision Under Risk. Econometrica 47, No. 2: 263–291

Kemp R (1985) Planning, Political Hearings, and the Politics of Discourse. In: Forester J (ed) Critical Theory and Public Life. MIT Press, Cambridge, MA

Knoepfel P, Weidner H (1982) Formulation and Implementation of Air Quality Control Programs: Patterns of Interest Consideration. Policy and Politics 10: 85–109

Kunreuther H, Slovic P (1996) Science, Values, and Risk. In: Annals of the American Academy of Political and Social Science, Special Issue. Kunreuther H, Slovic P (eds) Challenges in Risk Assessment and Risk Management. Sage, Beverly Hills and Thousand Oaks, pp 116–125

Kweit MG, Kweit RW (1981) Implementing Citizen Participation in a Bureaucratic Society. Praeger, New York

Lindbloom C (1977) Politics and Markets: The World's Political-Economic Systems. Basic Books, New York

Lindbloom C (1959) The Science of Muddling Through. Public Administration Review 19: 79–99

Lohmar U (1967) Wissenschaftsförderung und Politikberatung. Bertelsmann, Gütersloh

Luhmann, N (1984) Soziale Systeme: Grundriß einer allgemeinen Theorie. Suhrkamp, Frankfurt am Main

Lynn FM (1986) The Interplay of Science and Values in Assessing and Regulating Environmental Risks. Science, Technology and Human Values 11, No. 2: 40–50

Majone G (1989) Evidence, Argument and Persuasion in the Policy Process. Yale University Press, New Haven and London

Margolis H (1996) Dealing with Risk. Why the Public and the Experts Disagree on Environmental Issues. University of Chicago Press, Chicago 1996

Mayntz R (1993) Policy Netzwerke und die Logik von Verhandlungssystemen. In: Heritier A (ed) Policy Analyse. Opladen, Westdeutscher Verlag, pp 38–56

Mayntz R, Scharpf FW (1973) Kriterien, Voraussetzungen und Einschränkungen aktiver Politik. In: Mayntz R, Scharpf FW (eds) Planungsorganisation. Die Diskussion um die Reform von Regierung und Verwaltung des Bundes. Piper, München, pp 115–145

McDaniels T (1996) The Structured Value Referendum: Eliciting Preferences for Environmental Policy Alternatives. Journal of Policy Analysis and Management 15, No. 2: 227–251

Mohr H (1994) Das Expertendilemma. In: Stifterverband für die deutsche Wissenschaft (ed) Selbstbilder und Fremdbilder der Chemie. Stifterverband, Essen, pp 194–209

Mohr H (1996) Das Expertendilemma. In: Nennen HU, Garbe D (eds) Das Expertendilemma. Zur Rolle wissenschaftlicher Gutachter in der öffentlichen Meinungsbildung. Springer, Berlin, pp 3–24

Nelkin D Technological Decisions and Democracy. Sage, Beverly Hills

Nelkin D, Pollak M (1980) Problems and Procedures in the Regulation of Technological Risk. In: Weiss CH, Burton AF (eds) Making Bureaucracies Work. Sage, Beverly Hills, pp 233–253

Nowotny H (1979) Kernenergie: Gefahr oder Notwendigkeit. Suhrkamp, Frankfurt am Main

Nowotny H, Eisikovic R (1990) Enstehung, Wahrnehmung und Umgang mit Risiken. Schweizerischer Wissenschaftsrat, Bern

Olson M (1984) Participatory Pluralism: Political Participation and Influence in the United States and Sweden. Nelson-Hall, Chicago

O'Riordan T, Wynne B (1987) Regulating Environmental Risks: A Comparative Perspective. In: Kleindorfer PR, Kunreuther HC (eds) Insuring and Managing Hazardous Risks: From Seveso to Bhopal and Beyond. Springer, Berlin, pp 389–410

Pidgeon NF (1997) The Limits to Safety? Calture, Politics, Learning and Manmade Disasters. Journal of Contingencies and Crisis Management 5, No. 1: 1–14

Pollak M (1985) Public Participation. In: Otway H, Peltu M (eds) Regulating Industrial Risk. Butterworths, London, pp 76–94

President's Council on Sustainable Development (1997) Lessons Learned from Collaborative Approaches. President's Council on Sustainable Development, Washington, D.C.

Primack J, von Hippel F (1974) Advice and Dissent: Scientists in the Political Arena. Basic Books, New York

Ravetz J (1989) The Merger of Knowledge with Power. Mansell, London

Renn O (1995) Style of Using Scientific Expertise: A Comparative Framework. Science and Public Policy 22: 147–156

Renn O (1998) The Role of Risk Communication and Public Dialogue for Improving Risk Management. Risk Decision and Policy 3, No. 1: 5–30

Renn, O (1999) Participative Technology Assessment: Meeting the Challenges of Uncertainty and Ambivalence. Futures Research Quarterly 15, No. 3: 81–97

Renn O, Webler Th (1998) Der kooperative Diskurs – Theoretische Grundlagen, Anforderungen, Möglichkeiten. In: Renn O, Kastenholz H, Schild P, Wilhelm U (eds) Abfallpolitik im kooperativen Diskurs. Bürgerbeteiligung bei der Standortsuche für eine Deponie im Kanton Aargau. Hochschulverlag ETH, Zürich, pp 3–103

Renn O, Webler Th, Rakel H, Dienel PC, Johnson BB (1993) Public Participation in Decision Making: A Three-Step-Procedure. Policy Sciences 26: 189–214

Rip A (1985) Experts in Public Arenas. In: Otway H, Peltu M (eds) Regulating Industrial Risk. Butterworths, London, pp 94–110

Rip A (1992) The Development of Restrictedness in the Sciences. In: Elias N, Martins H, Wihtley R (eds) Scientific Establishments and Hierarchies. Kluwer, Dordrecht and Boston, pp 219–238

Rittel HWJ in cooperation with Webber MM (1992) Dilemmas in einer allgemeinen Theorie der Planung. In: Rittel HWJ (ed) Planen, Entwerfen, Design. Ausgewählte Schriften zu Theorie und Methodik. Kohlhammer, Stuttgart, pp 13–35

Rowe G, Frewer LJ (2000) Public Participation Methods: A Framework for Evaluation. Science, Technology & Human Values 225, No. 1: 3–29

Sager T (1994) Communicative Planning Theory. Aldershot, Avebury

Scarcinelli U (1990) Auf dem Weg in eine kommunikative Demokratie? Demokratische Streitkultur als Element politischer Kultur. In: Scarcinelli U (ed) Demokratische Streitkultur. Theoretische Grundpositionen und Handlungsalternativen in Politikfeldern. Westdeutscher Verlag, Opladen, pp 29–51

Scharpf FW (1991) Die Handlungsfähigkeit des Staates am Ende des zwanzigsten Jahrhunderts. Politische Vierteljahresschrift 32, No. 4: 621–634

Schimank U (1992) Spezifische Interessenkonsense trotz generellem Orientierungsdissens. In: Giegel HJ (ed) Kommunikation und Konsens in modernen Gesellschaften. Suhrkamp, Frankfurt am Main, pp 236–275

Sclove R (1995) Democracy and Technology. Guilford Press, New York

Stern PC, Fineberg V (1996) Understanding Risk: Informing Decisions in a Democratic Society. National Research Council, Committee on Risk Characterization. National Academy Press, Washington, D.C.

Susskind LE, Cruishank J (1987) Breaking the Impasse: Consensual Approaches to Resolving Public Disputes. Basic Books, New York

von Schomberg R (1992) Argumentation im Kontext wissenschaftlicher Kontroversen. In: Apel KO, Kettener M (eds) Zur Anwendung der Diskursethik in Politik, Recht, Wissenschaft. Suhrkamp, Frankfurt am Main, pp 260–277

von Schomberg R (1995) The Erosion of the Valuespheres. The Ways in which Society Copes with Scientific, Moral and Ethical Uncertainty. In: von Schomberg R (ed) Contested Technology. Ethics, Risk and Public Debate. International Centre for Human and Public Affairs, Tilburg, pp 13–28

Wachlin KD, Renn O (1999) Diskurse an der Akademie für TA in Baden-Württemberg: Verständigung, Abwägung, Gestaltung, Vermittlung. In: Bröchler S, Simonis G, Sundermann K (eds) Handbuch Technikfolgenabschätzung, Vol. 2. Sigma, Berlin, pp 713–722

Webler Th (1995) 'Right' Discourse in Citizen Participation. An Evaluative Yardstick. In: Renn O, Webler Th, Wiedemann P (eds) Fairness and Competence in Citizen Participation. Evaluating New Models for Environmental Discourse. Kluwer, Dordrecht and Boston, pp 35–86

Webler Th, Renn O (1995) A Brief Primer on Participation: Philosophy and Practice. In: Renn O, Webler Th, Wiedemann P (eds) Fairness and Competence in Citizen Participation. Evaluating New Models for Environmental Discourse. Kluwer, Dordrecht and Boston, pp 17–34

Weingart P (1979) Das 'Harrisburg-Syndrom' oder die De-Professionalisierung der Experten, Preface to Nowotny H (ed) Kernenergie: Gefahr oder Notwendigkeit. Suhrkamp, Frankfurt am Main, pp 9–17

Weingart P (1983) Verwissenschaftlichung der Gesellschaft – Politisierung der Wissenschaft. Zeitschrift für Soziologie 12: 225–241

Weinrich H (1972) System, Diskurs, Didaktik und die Diktatur des Sitzfleisches. Merkur 8: 801–812

Willke H (1995) Systemtheorie III. Steuerungstheorie. UTB Fischer, Stuttgart und Jena

Wynne B (1989) Sheepfarming after Chernobyl. Environment 31: 11–15, 33–39

Zilleßen H (1993) Die Modernisierung der Demokratie im Zeichen der Umweltpolitik. In: Zilleßen H, Dienel PC, Strubelt W (eds) Die Modernisierung der Demokratie. Westdeutscher Verlag: Opladen, pp 17–39

IV Case Studies

10 The Ethics of Technology Assessment in Health Care

Albert J. Jovell

10.1
Health Technology Assessment

10.1.1
The definition

A health technology is any kind of intervention used in health care for the screening, diagnosis, prevention, and treatment of diseases. The definition encompasses a wide variety of interventions including drugs or ways to organize the provision of health care, such as major ambulatory surgery or a specific software for a better planning of health care delivery. The amount of innovations coming from research and development processes in health care and the need to guarantee a high quality and efficient health care have urged the development of Health Technology Assessment. Also, it has promoted the development of specific institutions designed for assessing health technologies under government funding. These institutions known in most of the countries as Health Technology Assessment agencies have developed a network known as the International Network of Agencies for Health Technology Assessment (www.inahta.org). Also there is an organization known as the International Society for Technology Assessment in Health Care (www.istahc.org) for those who are interested in the issue.

There are several definitions of Health Technology Assessment in health care. All these definitions deal with the idea of assessing the impact of specific technologies on public health in terms of their effect and efficiency. Health Technology Assessment also includes evaluations on the impact of health technologies on the structure, process and outcomes of health care. The results of these assessments help policymakers in making decisions on the financing and use of the assessed technologies. Also it might help other stakeholders rather than government, such as physicians, consumers, patients, industry and third-payers to make decisions on the adoption and use of health technologies.

A more fashionable definition compares Health Technology Assessment with the concepts of evidence-based medicine and evidence-based health care.

All these concepts integrate the idea of using research data to aid decision-making in health care. In this chapter every time we refer to Health Technology Assessment we will refer also to the evidence-based medicine and health care as synonymous concepts. From a philosophical perspective, decisions founded on data allow decision-makers to be made accountable for their actions.

10.1.2
The method

Alongside different related definitions of Health Technology Assessment, there are also different methodological approaches for its implementation. One common trend to all these methodologies is to put emphasis on the method rather than the person who makes the recommendations. Traditional ways of making recommendations in health care such us consensus opinion or expert opinion should no longer be valid under a Health Technology Assessment approach. Health Technology Assessment intends to overcome uncertainty in decision making through formal methods of producing and synthesizing data. In doing so, they follow a major rule of kantian philosophy which moves from subjective validity to objective validity. Thus, the Health Technology Assessment process can be defined as a knowledge driven process for better decision making in health care.

Table 10.1. Systematic review of evidence

1. Search for direct and indirect evidence to answer a specific policy question.
2. Presentation and classification of the evidence according to specific selection criteria.
3. Synthesis of the direct evidence: meta-analysis
4. Integration of the evidence:
   - Decision analysis
   - Cross-national comparison analysis
   - Economic analysis
   - Appropriateness studies
   - Ethical analysis
   - Legal analysis
5. Evidence-based policy recommendations:
   - Clinical practice guidelines
   - Policy statements
   - Public health policies

In Table 10.1 we describe a process for formal Health Technology Assessment. The concept might be known also as systematic review of the evidence. It is a step-by-step process designed to produce knowledge. The first step includes searching for so-called direct evidence. This is the evidence produced by formal research studies assessing the health effect of the technology. In a specific context of care, it might include the search for indirect evidence.

This is the evidence produced by health data that have not been collected for the purpose of a study, such as clinical records data or accounting data. In a second step, direct evidence is classified according to its internal validity resulting in what is known as level of evidence scales (Table 10.2 and Table 10.3). Levels of evidence scales are useful to make recommendations according to the strength of evidence (Table 10.4). Internal validity refers to the robustness of a study design as a method to provide results at the highest strength, while external validity puts the emphasis on the robustness of a study to extrapolate study results to different contexts of care. If we should find several randomized controlled trials we have to develop an evidence table to summarize studies according to subject and design features. A third step would be, if possible, to pool data from different studies sharing the same study design and testing the same hypothesis. Meta-analysis is a method to pool data from randomized controlled trials that are clinically combinable. It might occur that in the process of Health Technology Assessment we might find some meta-analysis published in the medical literature. On the other hand, in other cases it might not be possible to pool data from several studies.

Table 10.2. Internal validity criteria for study design

- Randomization
- Presence of a control group
- Longitudinal study
- Prospective follow-up
- Blinding of study subjects and researchers to treatment allocation

Once we have single or pooled data on the health effect of the technology it might be possible to combine them with indirect evidence. The fourth step is the process of integrating direct and indirect evidence that helps to put the results of the assessment in context of care. This step also includes the possibility to individualize the results of the assessment according to patients' characteristics through decision analysis or to assess the global impact of the technology through economic analysis and related methodologies. Finally, the results obtained in the five-step process yield the production of explicit recommendations that can be clinical practice guidelines or public health policies. An example of clinical practice guideline for the treatment of Helicobacter pilory infection included all of the above-described steps (Jovell 1998). The guideline recommended the prescription of triple therapy composed by two antibiotics and omeprazole as the most efficacious disease treatment. This recommendation came from a meta-analysis of existing studies. Also the guideline recommended empirical therapy in the treatment of recurrence ulcer as the most efficient treatment as a result of a cost-effectiveness analysis.

Table 10.3. Levels of evidence scale

Study design	Quality of the evidence	Strength of recommendations
Meta-analysis of randomized controlled trials Randomized controlled trials (enough statistical power/prevalent disease) Randomized controlled trials (low statistical power/low prevalent disease)	Good	A
Randomized controlled trial (low statistical data/ prevalent disease) Non-randomized controlled trial Cohort study Case-control study	Fair	B
Clinical series Consensus conference Expert opinion Single case/anecdote	Bad	C

Table 10.4. Strength of recommendations

A: There is good evidence to support the recommendation that the technology should be adopted or excluded in the delivery of health care.
B: There is fair evidence to support the recommendation that the technology should be adopted or excluded in the delivery of health care.
C: There is insufficient evidence to support the recommendation that the technology should be adopted or excluded in the delivery of health care.

10.1.3
The health outcomes

Health Technology Assessment is a process that has no single perfect methodology. The method of choice depends on the policy question in search of an answer. Therefore, the five-step process described in a former section can be modified to adapt a specific study design to answer concrete research questions. This can be the case when an economic analysis is carried out to answer a question on the efficiency of adopting a specific technology, such as whether to choose streptokinase or rt-PA in the treatment of acute myocardial infarction (The GUSTO investigators 1993).

In most of the cases a process of Health Technology Assessment is carried out to assess the global impact of a technology following the five-step process already described in this chapter. In this situation there is a rationale for outcome selection criteria (Table 10.5). The rationale indicates that the first criteria for assessment should be the measure of the efficacy of an intervention. Efficacy is the measure of the effect of a health technology under

Table 10.5. Health outcomes ranking criteria for Health Technology Assessment

Outcome	Study design	Type of data	Validity
Efficacy	Randomized controlled trials	Experimental	Internal
	Meta-analysis	Experimental	Internal?
Effectiveness	Observational studies	Context of care	External
Cost	Accounting data	Context of care	External
	Data collection and evaluation		
Cost-effectiveness	Economic analysis	Context of care	External
Appropriateness	Observational studies	Context of care	External
Need	Observational studies	Context of care	External
	Economic analysis		
	Case-studies		
Ethical impact	Case-studies	Cultural context	External

an experimental design, mostly a randomized controlled trial (Jadad 1998). Once we have obtained data at the highest level of internal validity using experimental designs such as randomized controlled trials or meta-analysis of that type of studies, we have to collect data on the real practice in which the technology is going to be applied. This type of data applies to the external validity of the assessment or its applicability to a specific group of patients and clinical settings. The first type of data to be collected is on effectiveness, being the measure of the effect of an intervention under standard conditions of clinical practice. Obviously, effectiveness is not the same as efficacy because of the differences between experimental conditions and real clinical practice. Thus, effectiveness should be lower than efficacy. An example of this difference can be seen in the treatment of elevated blood pressure, which is one of the most prevalent risk factors for disease. The experimental model uses randomized controlled trials to compare an experimental drug with placebo assessing as outcome the number of vascular events avoided after some years of therapy. If the study results showed that the drug prevented a higher number of events than placebo and this number is statistically significant, we can conclude that the efficacy of the drug is higher than placebo and recommend it for medical practice. Once, the drug is prescribed in medical practice it might happen that it does not have the expected level of therapeutic effect because patients are not compliant with the prescribed treatment due to drug side-effects. Therefore, patients' non-compliance might be the reason why the expected efficacy obtained in randomized controlled trials does not correspond to the level of effectiveness measured in non-experimental "real life" clinical practice.

Another criteria to assess is the cost of the intervention and, what is most important for resources allocation decisions, the expected effectiveness gained by an increase in the level of investment. This last measurement is a measure

Table 10.6. Economic analysis methods

Methods	Health outcomes unit	Cost unit	Ratio
Cost analysis	none	currency	none
Cost-benefit analysis	currency	currency	savings
Cost-effectiveness analysis	health effect	currency	cost-effectiveness ratio
Cost-utility	utility	currency	QALY's DALY's

of the efficiency of the interventions and is made through economic analysis methodology (Table 10.6). The description of all these methods is beyond the scope of this chapter but there are three considerations that should be taken into account when considering the applicability of economic analysis in health care (U.S. Department of Health and Human Services 1996). The first of them refers to the use of cost-analysis as the most published economic method in the health care literature. Cost-analysis might be an incomplete method for assessing health technologies because it focuses on the cost of the technology without any consideration of its effects. In the field of Health Technology Assessment, what is important for decision making is to be able to assess what kind of improvement is expected at the different cost levels. A second consideration refers to the adoption of cost-benefit analysis, a method that measures health effects in monetary terms. It is very difficult at the ethical level of thinking to find an acceptable method to measure human life in monetary terms. Therefore, cost-benefit analysis does not seem to be a reasonable method for health care assessment because it attempts to convert subjective values, such as human life or health improvement, into monetary values. Despite the existence of some specific techniques to transform health effects in monetary terms, such as the willingness to pay technique, there is no consensus on the fair application of these methods in Health Technology Assessment. Finally, a third consideration refers to the concept of utility which takes into account the health effect as it was measured and the value placed on that effect by individuals. Therefore, quality-adjusted years of life and disability-adjusted years of life are units that integrate the true effect of the intervention and the value to be placed on it. The cost-utility ratios allow decision-makers to compare the efficiency of different intervention across health care through a common metrics. Criticisms of cost-utility analysis come from its methodological weakness. In this sense, there are two major criticisms against this method. The first of them refers to how to deal with the different values that an individual might have depending upon his or her state of health. The question to be answered is whether values placed by an individual on a theoretical improvement in health care will differ according to whether he has the disease or is healthy. The second criticism refers to the ethical objections that might arise from extrapolating population utility values to a single individual patient.

The above mentioned comments on the limitations of economic analysis might not discourage its use in Health Technology Assessment. One of the basic considerations to bear in mind in the evaluation of medical and health interventions is the absence of a perfect method of assessment. On the other hand, any of the interventions in need of an assessment produce costs and effects that need to be compared with other alternatives in order to make a fair decision. To do so the most suitable technique for economic analysis might be the cost-effectiveness analysis. This method compares the effect and cost of an intervention with another one or a do-nothing alternative. The results of the analysis provide a measurement of the expected improvement units in terms of health effects attributable to the intervention as they relate to the different units of cost. Therefore, the cost-effectiveness ratio might be expressed as the cost per year of life gained or the cost per day of disability saved, among other types of ratios. Cost-effectiveness analysis based recommendations are straightforward when the results of the analysis of comparing two different technologies indicate that at the same cost one of them produces a higher effect or, conversely, for the same effect one of them is cheaper. The ethical question arises when one of the technologies produces an improvement in health effects at a much higher cost. In this situation decision-makers have to confront the decision of adopting a technology in which its increasing cost is much higher that the increasing effect it produces. This is the case of whether to prescribe streptokinase or rt-PA for the treatment of acute myocardial infarction. A randomized controlled trial called GUSTO indicated that rt-PA reduced death rate after acute myocardial infarction 1 percent lower compared with streptokinase (The GUSTO investigators 1993). On the other hand, the cost of rt-PA was 12 to 20 times higher than streptokinase posing this situation the challenging question of having to decide whether the 1 percent death reduction justifies a higher cost (Mark et al. 1995). In addition, a second question arises when the results obtained in a randomized controlled trial have to be extrapolated to specific contexts of care for the treatment of acute myocardial infarction.

Finally, data on appropriateness might be useful to make decisions on the appropriate prescription and the appropriate use of a health technology. In the case of high blood pressure, data from randomized controlled trials indicated which level of blood pressure might benefit from the therapy and, therefore, it was appropriate to prescribe the treatment. In addition, a measure of compliance with the therapy might indicate if patients with high blood pressure follow the prescribed treatment. Another criteria for the evaluation of health technologies are based on the assessment of health needs. The concept of need refers to epidemiological, economic and social criteria. Difficulties in assessing the need for a particular health technology might discourage its funding by third payers. That was the case with Viagra, a drug for the treatment of impotence, which is not funded by public health care systems because there was no consensus on how to define when and for which type of patients it is appropriate.

10.2
Health systems in transition

10.2.1
Social change in health care

Looking at the context of health care, we can see that it has changed shape dramatically. The emerging social changes move health care systems to confront new dilemmas. These changes might be identified through the so-called health care transitions (Table 10.7). Major demographic changes caused by increasing life expectancy and migratory movements produce new and more health needs and, thereby, an increasing demand for health services. In addition, epidemiological change results in an increasing pattern in the prevalence of chronic diseases and comorbidities. The financial sustainability of health care systems is not only challenged by major changes at the demographic and epidemiological levels but also by the increased rate of innovations from research and development processes. New and expensive technologies are produced mostly without a proper assessment of their need, efficacy, effectiveness, cost-effectiveness, and appropriateness. Despite the lack of a proper assessment new discoveries in the health field increase public expectancies on the possibilities of health care.

Table 10.7. Healthcare transitions

- Demographic
- Epidemiological
- Economic
- Technological
- Labour market
- Judiciary
- Mass-media
- Political
- Ethical

Innovations in the field of communication and information technologies are changing current patterns of the delivery of health care through the adoption of the new electronic medicine paradigm. Also they contribute to modify public demands on health care by means of up-to-date, free, and universal access to health information from any place in the world via the Internet. The so-called health information explosion on the Internet has a comparatively strong impact on the growing mass-media interest in health care. This interest has its main focus on the problems associated with health care following the principle of "good news, no news" and in potential breakthrough discoveries. In this sense, mass-media focus on health is mainly targeting the unusual and new rather than the success of the day-to-day provision of health care. Public expectancies and the mass-media impact on health news result in an increasing number of claims on the legal system. Additionally the information explosion phenomena exceeds the physician's ability to cope with all available information. Moreover, most of the available health information is redundant, low quality and needs to be updated.

Alongside the changing health information trends, an economic transition following globalization changes will affect the way health care is organized and provided. Public funding might be maintained for the coverage of a basic health care package in most developed countries. The provision of health care can be affected by a mixed system of private and public corporations under different and specific co-payment arrangements. Priority setting will be a major trend in the definition of the limits of public funding and system sustainability. The health care industry will be organized around major corporations. Corporations might be diversified and big so as to promote economies of scale. Also, small companies will be organized around highly skilled workers and specialized knowledge.

All these transitions can result in major changes in the relationship between consumers and systems of care. These changes might include the promotion of the right to know as a major ethical duty, the democratization of information, and the establishment of consumer empowerment policies. As a consequence of these changes and the increasing public expectancies towards the possibilities of health technologies, the traditional model of physician-patient relationship based on the asymmetry of knowledge might be distorted. These scenarios following health transitions yield a system in which the definitions of health and health care are going to change dramatically. This situation poses a last challenge: can we change or modify people's values? In other words, should we promote an ethical transition for health care?

10.2.2
Challenges for Health Technology Assessment

Health Technology Assessment faces major challenges due to the lack of a proper definition of need and coverage in health care, the weight of inappropriate research in decision-making, the uncertainty associated with medical decisions, the limitations of efficacy data when it is transferred to different populations and clinical settings, and the conflicting values among different stakeholders and their different definitions of social justice.

All the changes briefly mentioned in the above section challenge directly the sustainability of health care systems. The fact that some of these changes are related to the need for health technologies or the availability of new technologies determines the emergence of Health Technology Assessment and evidence-based medicine to yield an accountable decision-making process in health care. One of the first challenges to the sustainability of health care would be represented by what I call the "coverage frontier". It represents the conflict among access, quality, and cost in health care. Systems of care can not delivery services that offer the highest quality for everybody without controlling cost. Then, we have to choose some of the health outcome criteria defined earlier in this chapter or we have to choose a combination of them if we want to be accountable and realistic in managing health care. In this sense, Health Technology Assessment might be of assistance in informing decision-making.

There are some questions that evidence-based medicine might help to answer, such as "why is this high level of uncertainty about the benefits or harm of most of the technologies – not only the new ones but also the ones we already have?". The issue of uncertainty is important because a doctor or a manager is confronted with more information in one year than he/she is able to read in all his/her life. In addition, medicine is most of the time based on presumptions and physio-pathological never-tested hypothesis. Therefore, levels of evidence scale might be useful for aiding health care professionals in clinical decision making. On the other hand, because of uncertainty and information overload there is a need to design a primary research project collaboration network covering all health care research. The network can allow us to get more value for money in the money we spend on research. This idea came after searching a bibliographic database in health care known as Medline for research on prostate cancer. Most of the research in this field during ten years – 1988–1997 –, about 63 per cent, was case series, which did not test any research hypothesis, and 32 per cent were reviews based on expert opinion. Only 5 per cent of these 43,140 bibliographic cites were randomized controlled trials. None of them was testing the hypothesis that would be of major public health interest such as whether or not prostate cancer screening increases patients' survival. At this moment, we do not have the answer to this question despite widely adopted prostate cancer screening. Also, there are at least 13 different therapies that have been proposed for the treatment of localized prostate cancer without any supporting high quality evidence. At the end, the prostate cancer lesson summarizes a common trend in medical research: too much research published, very little quality.

Despite the presence of uncertainty there is some evidence on some of the decisions we make in health care. As we said before, major evidence comes from randomized controlled trials but there are also other data, which we might call the "grey data". It would include data that we really do not know how to deal with but it exists and constitutes a merge of local data and low quality evidence. Moreover, most of the decisions might be placed in a "black box", or an area in which we know there is a need to make decisions but we lack data to support them. Again, quoting the same case of prostate cancer treatment. At this moment, we do not know the difference in terms of survival using the most aggressive available treatment – radical prostatectomy – or using the most conservative treatment – watchful observation. Current data are controversial but what is known is that patients' quality of life might be affected if we choose a very aggressive treatment.

One possible approach to overcome uncertainty and help decision-makers in their tasks might be educating them on the basic steps and goals of Health Technology Assessment. Education will help them go through difficult situations. The case of prostate cancer screening is also illustrative of the different approach to the adoption of a health technology depending on whether or not physicians have been educated in Health Technology Assessment. Three of the most important medical societies in the United States – the American Cancer Society, the American Association of Urology, and the American College of Radiology – advocate screening for prostate cancer based on the digital rectal examination and the prostate serum antigen every year for males older than 50 years (The Clinician's Handbook 1994). On the other hand, we have prestigious medical societies in the United States and Canada who will discourage this practice so are governments in countries like United Kingdom, Sweden, and Canada that also really do not recommend this practice for screening. It would be very difficult for a practitioner to make decisions on prostate cancer screening under conflicting recommendations. Therefore, an evidence-based health care approach would indicate that there is not evidence supporting a higher life expectancy attributable to the benefits of prostate cancer screening and therapy. Paradoxically, despite this lack of evidence as we said before there is a growing number of treatments for prostate cancer.

Other challenge for Health Technology Assessment will be the need to define how much evidence can be qualified as enough evidence for effective clinical decision making. In a meta-analysis on angioplasty published in (Weaver et al. 1997), it is indicated that angioplasty reduced hospital mortality compared with thrombolytics after an acute myocardial infarction. However, this answer based on therapeutic efficacy might not be reproducible in terms of effectiveness because you need a very high expertise on training and ability to do angioplasties. Skilled cardiovascular surgeons are a very expensive resource alongside special cardiovascular intensive care units. In addition, angioplasty is an evolving technology easily replaced for theoretically improved technologies such as angioplasty plus stent. Therefore, even if we have this information and have performed the meta-analysis, there are a lot of things we do not know and many questions we need to answer before adopting a new procedure. This is also common to all treatments for coronary heart disease. There is a big difference in terms of cost, physicians' and patients' preferences, and also in terms of needs regarding angioplasty, thrombolytics, by-pass surgery, transmiocardial laser, atherectomy, and coronary stent. Moreover, there is no system of care that can provide the most costly treatment for everybody who has a myocardial infarction.

In addition to the coverage limits of health care, there are two new frontiers that might benefit from Health Technology Assessment. The first of them is the decision-maker frontier. It represents conflicting values among all the health care stakeholders. Patients, consumers, physicians, managers, government, and other stakeholders might have different values placed around the same decision-making process. Physicians and patients might be focused on the principle of beneficence when assessing the need for a new treatment while managers might be discouraged by its high cost and government by the problems of universal accessibility produced by its adoption. Health Technology Assessment might help to put in a population perspective the data required to weight the potential benefit of the new treatment against its cost both for a single patient and for all patients who needed it. The second frontier is the moral frontier of possibilities. It indicates the conflict among major social justice theories such as utilitarian, liberal, and egalitarian ones. Health Technology Assessment might provide the data needed for a more explicit definition of fairness and justice in the distribution of health resources.

10.3
Knowledge management

10.3.1
The knowledge based system of care

The Health Technology Assessment and the evidence-based health care approaches described above refer basically to what may be called explicit knowledge. This type of knowledge is mainly quantitative and easy to share among different interested parties. It can be global when it is coming from meta-analysis and published research and it can be local when it is coming from the quantitative assessment of introducing and adopting health technologies in a specific context of care. Despite the growing importance of explicit knowledge, there is another kind of knowledge that might be called tacit knowledge that seems to play a major role in decision-making. Tacit knowledge refers to individual and group values, individual expertise, collective interaction, and ideas. It is mainly qualitative. Therefore, when analyzing decision-making in health care it is important to consider both the explicit and the tacit knowledge. Some of the case-studies described in this chapter, such as the prostate cancer screening indicated the different weight given in the decision-making process to both types of knowledge. Whether to screen for prostate cancer or not is weighted differently by those who based their decision on explicit knowledge compared to those who relied on tacit knowledge. Those who base their decision on tacit knowledge value prostate cancer as a life threatening disease in need of treatment.

The two different types of knowledge might be a source of latent moral conflicts in decision-making. Most of these conflicts concern the goals of health care. The prostate cancer screening dilemma poses some of these moral conflicts that can be extended to other health issues. The first moral conflict is whether health care should be oriented to check for diseases or for good health. The question here is: should it be a disease screening policy searching for disease at early stages or should it be aimed at maintaining a good health status? The second moral conflict is whether health care should be focused on the disease or on the patient. If it is focused on patients' care then their preferences should be elicited and respected despite Health Technology Assessment recommendations that were focused on the technology and the disease. The third moral conflict is whether health care should promote an increasing life expectancy or an improvement in quality of life. In the case of prostate cancer the question would be whether to choose unproven therapies intending to cure the disease for cancer patients although they may have important side-effects that affects patients' quality of life.

10.3.2
Dissemination of knowledge

The process of evidence-based medicine really begins with a question that comes from a health problem. If we are lucky and we have the evidence, then the next issue will be how we can implement the evidence. To implement the evidence, first of all, we have to put the evidence into the hands of the people who are going to use it. Dissemination is a set of customized actions designed to promote a behavioural change, linking knowledge with practice in those who make decisions in health care. Dissemination plays a key role in the Health Technology Assessment and evidence-based medical processes. Also, we have to assess the impact of implementing evidence-based medicine. To assess it, we have to formulate another research policy question. So this is a cyclic process. To implement evidence-based medicine, we need basically to have the evidence, to disseminate, to market, and to assess its implementation.

Dissemination is a different concept from diffusion because diffusion is a passive action and dissemination is an active action. Dissemination is focused on behavioural change and on how to convince the people who make the decisions to adopt this evidence in their practice. The key elements of a dissemination strategy will be the accessibility of information, access to the evidence, and the education and the training of professionals to use the evidence. It would be useful to compare what the evidence is telling us about how we should do things and how the things are being done. Thus, information, and knowledge if possible, would be needed on current practice and care procedures to help to go through the dissemination process. Also, dissemination can involve patients. It would be useful to inform them about what the evidence is telling us about the treatment or care of the disease and, henceforth, eliciting their preferences. In this sense, dissemination involves considering the different stakeholders of the system and their values, which may be so diverse that they can interpret the evidence with different meanings. So, we have to disseminate to a specific target group. Some of the stakeholders that constitute specific dissemination targets are policy-makers, citizens, mass media, and the judiciary. All of them play a major role in health care. Mass media might really promote the adoption of technologies that has no evidence of effect. Also the judiciary is very focused on personal stories rather than on the average population data. Management might have a big role in promoting the use and adoption of evidence for rationing.

These are some key elements needed to adopt a knowledge-based system of care. The first one is to establish proper working arrangements to use the evidence in practice. The second element is to inform those who are going to use this information for the care of patients about what is effective and what is not. The third element is to finance any initiative that is going to produce the evidence needed for making decisions. The fourth element is to improve the healthcare professionals' skills to use the evidence and train them to understand and interpret it. The fifth element is to put the money and the incentives for people to apply or implement effective treatments. The sixth element is to look for co-ordination between the different levels of care: primary care and hospital care; outpatient care, secondary care and tertiary care. In the hospital, we also have to look for co-ordination between nurses, physicians and managers. Finally, the seventh element is to promote initiatives producing tacit knowledge for improving the effectiveness of the system of care.

The production of tacit knowledge plays a key role to put Health Technology Assessment into practice. The assessment of clinical practices at local level also has a key role to play in providing feedback to physicians and nurses on how they can improve their practice. This approach of producing local data and knowledge is important to promote change because some professionals still rely on authority while others rely a lot on unproven assumptions. In addition, some rely on experience and others on co-operation, consensus or scientific evidence. In this sense, decision-making has its own transition moving from more subjective ways of thinking to more objective ones based on information and evidence. On the other hand, dissemination faces important barriers to go forward with its aims. One of them is the high prevalence of the traditional model of care based on experience and an authoritarian point of view. Also, the organization of care is not always adequately organized to promote the use of evidence. Obviously there are conflicts of interest in health care like in any other dimension of life. Professional values play a role. Other factors include the lack of incentives to use evidence and the absence of necessary knowledge and skills. Health care professionals that have not got adequate training in Health Technology Assessment may feel themselves insecure when we are giving them data in a very sophisticated way.

Finally, it is important to assess the impact of our interventions. However, to assess the impact of implementing evidence-based medicine, one should assess the effectiveness of the dissemination process. Specific research should look for any change produced in the system and you should be able to link evidence with practice and to look for improvement or evaluation in health outcomes. Despite, the need to assess the impact of Health Technology Assessment we have to keep in mind that we can not apply evidence to every aspect of the decision-making process. It is very difficult to obtain data on all health interventions in terms of indicators such as efficacy, effectiveness, efficiency, equity, accessibility, appropriateness and need. Therefore some of the decisions are going to be made under conditions of uncertainty. What we can do is to develop a more reliable framework for making decisions depending on the level of evidence that we have. If the evidence is conclusive like in the case of the thrombolytic therapy in the treatment of acute myocardial infection, a clinical practice guideline for implementation can be developed. Evidence is highly conclusive that thrombolytic therapy can save lives (Lau 1992). If there is conflicting interpretation of evidence then maybe we should go for having data on another type of indicator such as acceptability. If available assessment showed contradictory outcomes like in the prenatal screening for Down's syndrome, then we have a real case for patient preferences and for value negotiation (Serra-Prat et al. 1998). And if there is no evidence, like in the case of prostate cancer screening then we have to do more good quality research.

10.3.3
Ethics and knowledge in health care

A specific case study in prenatal screening for Down's syndrome might be helpful to ascertain the complexities of the practice of Health Technology Assessment and evidence-based health care (Serra-Prat et al. 1998). We looked at the evidence and all the data relating to the diagnostic tests of biochemical and ultrasonography used for diagnosing whether the fetus of a pregnant woman had Down's syndrome. We identified eight different policies and applied them to the context of care in Catalonia using both published and local data. We found that detection rates were higher on those policies that had the higher cost and the higher number of spontaneous abortion attributable to amniocentesis. It was not possible to propose Down's a policy based on evidence for the screening of Down's syndrome as the best policy. The evidence obtained does not support all the outcomes in one direction. It supports one outcome in one direction and the others in the opposite direction. When you have all this information, you might decide whether to choose a policy in which you have maximized detection of Down syndrome or choose a policy in which you minimized the risk of spontaneous abortion. If you minimize risk, you have a higher probability of a false negative result but you avoid fetal losses due to amniocentesis. If you have a higher detection rate, you maximize detection but you are going to have a higher number of fetal losses due to invasive procedures. In case of conflicting evidence, patient preferences and value negotiation might play a big role in the trade-off between potential risks and benefits.

The case of screening for Down's syndrome illustrates one of the major ethical challenges for health care decision-making: what to do in the case of conflicting evidence? In this specific case evidence-based information can be a source for public deliberation if a public health decision should be made or for physician-patient deliberation if an individual decision should be made. Evidence-based deliberation might be helpful to overcome what might be called the moral conflicts or moral disagreements in healthcare (Guttmann et al. 1996). Also, deliberation might be helpful to confront the above mentioned frontiers. One of them refers to who is going to make the decision. It is a fashion now to say "patient preferences" must be taken into account. However, if patients prefer the most expensive treatment, who is going to pay for that? Evidence-based deliberation is going to help to define what is now called shared decision-making. Thus, in the case of prostate cancer screening and Down's syndrome screening it should be weighted what level of decision should be made by consumers, by policy makers or by health professionals. On the other hand, there are a lot of ethical implications defined by the social justice or moral frontier of possibilities. Policy-makers usually talk about obtaining the highest value for the invested money when allocating health resources. Therefore, if they were utilitarian they made policies looking for efficiency and cost-effectiveness. On the other hand, policy-makers would also like to respect freedom of choice, either a doctor's autonomy or a patient's autonomy. Most of welfare States had what we call an "egalitarian system for care". We are now also facing what we call the moral frontier of possibility in order to define which will be the values that best fit our system: utilitarian, liberal or egalitarian.

The moral frontier leads to a very important question: why do we focus on evidence of the effect of an intervention and ignore all potential knowledge on values? Even if there is evidence of the true effect of the intervention, on population needs and of the costs involved, most of the decisions we make in health care are based on values. The clinical relevance of measuring the effect of an intervention can be interpreted differently just like the way needs and costs can be interpreted differently, too. Therefore, some kind of research on values is needed, Public deliberation on values can, therefore, be a way to include Health Technology Assessment in an appropriate manner into clinical practice and public health policy. To achieve this, public deliberation should go forward through the constitution of partnerships among the different stakeholders. These would include: therapeutic partnerships between physicians and patients, professional partnerships within the different health professions, corporate partnerships between professionals and third-party players, and social partnerships between professionals and other stakeholders such as journalists or governments. By this means the knowledge-based deliberation process could be put into practice. Thus the present and the future of Health

Technology Assessment and evidence-based medicine depend on the moral deliberations concerning the definition of health and the aims of health care systems.

References

Guttmann A, Thompson D (1996) Democracy and disagreement. The Belknap of Harvard University Press, Cambridge, MA

Jadad A (1998) Randomised controlled trials. BMJ books, London

Jovell AJ, Aymerich M, Garcia-Altes A, Serra-Prat M (1998) Guía de práctica clínica para el tratamiento erradicador de la infección por Helicobacter pylori en atención primaria. Agencia de Evaluación de Tecnología Médica, Departamento de Sanidad y Seguridad Social. Barcelona (see English version in www.aatm.es)

Mark DB, Hlatky MA, Califf RM, Naylor D, Lee KL, Amstrong PW, et al. (1995) Cost-effectiveness of thrombolytic therapy with tissues plasminogen activator compared with streptokinase for acute myocardial infarction. The New England Journal of Medicine 332:1418–24

Lau J, Antman EM, Jimenez-Silva J, Kupelnick B, Mosteller F, Chalmers TC (1992) Cumulative meta-analysis of therapeutic trials for myocardial infarction. The New England Journal of Medicine 327:248–54

Serra-Prat M, Gallo P, Jovell AJ, Aymerich M, Estrada MD (1998) Trade-offs in prenatal detection of Down syndrome. American Journal of Public Health 88:551–7

The Clinician's Handbook of Preventive Services (1994) Office of Disease Prevention. US Department of Health and Human Services. International Medical Publishing, Inc, Washington, DC

The GUSTO investigators (1993) An international randomized trial comparing four thormbolytic strategies for acute myocardial infarction. The New England Journal of Medicine 329:73–82

U.S. Department of Health and Human Services (1996) Cost-effectiveness in health and medicine Report to the U.S. Public Health Service by the Panel on Cost-Effectiveness in Health and Medicine. US Government Printing Office, Washington, DC

Weaver WD, Sims RJ, Betriu A, Grines CL, Zojstra F, Garcia E, et al. (1997) Comparison of primary angioplasty and intravenous thrombolytic therapy for acute myocardial infarction: a quantitative review. JAMA 278: 2093–8

11 Transition Management: a promising policy perspective

Jan Rotmans, René Kemp, and Marjolein van Asselt

Abstract

This chapter reflects the scientific examination of the concept of transition from an integrated perspective. It develops a model for working towards a transition in an iterative, adaptive manner, based on a philosophy of learning-by-doing and doing-by-learning. The model is applied to the case of a low emission energy supply in the Netherlands. Transition management offers a long term vision and framework for decision-making that aims to achieve greater coherence within public policies and between public policy and private action. Through transition management society's problem solving capabilities are mobilised and translated into a transition programme that is legitimised through the political process. It is a promising policy perspective that has been adopted by Dutch environmental authorities to deal with problems that require structural change.

11.1
Origin and nature of the concept of transition

The concept of transition has its roots in biology and population dynamics (Davis 1945; Notestein 1945). In the field of social developments, the transition concept was empirically founded and validated using *demographic transition*. The birth and death rates are high and in balance with each other during the predevelopment phase of demographic transition (see figure 11.1). During the take-off phase, the death rate falls, mainly as a result of improvements in hygiene and health care. During the acceleration phase, the death rate falls spectacularly, whilst the average birth rate remains high. This causes the population to increase at a fast rate. During the stabilisation phase, the dominant determinant is a fall in the birth rate, mainly caused by *innovation*, a complex social process whose driving forces are education, labour participation of women, economic development and family planning.

In general, a demographic transition covers a period of at least a generation, although it may vary from approximately 25 years to hundreds of years. In a fully successful demographic transition, a new balance is created with low birth and death rates, but with a considerably larger population size than at the beginning. In a failed demographic transition, on the other hand, the birth rate does not fall to the same extent as the death rate. As a result, the transition does not come to a new balance and the size of the population keeps increasing significantly. Demographic transition has become a historical fact in approximately 30 countries, including all the industrialized countries of Europe, Japan and North America. In many developing countries, however, the demographic transition must still come to full bloom, whilst there is still the risk that it may come to a standstill.

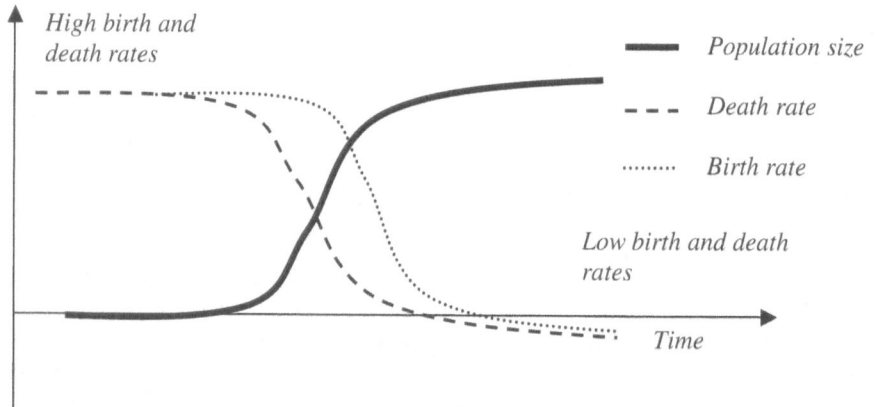

Fig. 11.1. Demographic transition

The example above of demographic transition shows that the concept of transition is definitely not something new. The concept of transition is also regularly used to interpret technological changes, which are otherwise called socio-technical transitions (Geels and Kemp 2000; Weaver et al. 2000). What is new, however, is the use of the concept of transitions to describe broad social changes and to explain their mutual connection. This implies that the concept of transition is used at a conceptual level to structure divergent social phenomenon. The concept of transition is, therefore, a heuristic, with which the complexity and coherence of societal changes can be described and explained. There is still not enough scientific proof to legitimize the use of the concept of transition in such a broad sense. So the hypothesis is that the concept of transition forms a suitable method to adequately describe complex societal dynamics. Nevertheless, there are sufficient indications that the concept of transition is an attractive, useful and helpful aid for figuring out social complexity and coherence (Ness et al. 1993; Rotmans 1994, 1995).

A transition can be defined as *a gradual continuous process of change where the structural character of society (or a complex sub-system of soci-*

ety) transforms. The description of demographic transition indicates that not every country follows the demographic transition curve and there are large differences in the period of time and the size of change between countries which have undergone demographic transition. The transition process is not deterministic, because during a process of change, adaption to, learning from and anticipating new situations will be needed. Transitions are, therefore, not patterns which determine what should happen. Transitions are not blueprints, but rather possible development paths where the direction, size and speed can be controlled through policy and specific circumstances.

A transition is the result of developments in different domains. In other words, a transition can be described as a set of connected changes, that reinforce each other and which take place in several different areas, such as technology, economy, institutions, behaviour, culture, ecology and world views/paradigms. Transitions are, therefore, a combination of changes in different domains, which develop through interaction with each other in a certain direction. A transition can be seen as a spiral that reinforces itself. In other words, there is multiple causality and co-evolution caused by interaction between developments in the various domains and by independent developments. A transition is, by definition, multi-dimensional and, therefore, has different dynamic layers. Even if the transition is emphatically supported by developments in a certain domain, it is always necessary for several developments in different domains to come together and reinforce each other. This also means that, in principle, all movements in the societal dynamics have an *impulse value* for transitions, or in other words, they can provide the *flywheel* force. At the same time, a transition cannot be caused as a result of only one innovation in only one domain. So transitions can be considered as a process of system innovation and social transformation that requires a combination of reinforcing developments in various areas.

At the conceptual level, we can distinguish four different transition phases (see figure 11.2):

1. A *predevelopment* phase of dynamic balance where the status quo does not visibly change.
2. A *take-off* phase where the process of change gets underway because the state of the system begins to shift.
3. An *acceleration* phase where visible structural changes take place through an accumulation of socio-cultural, economic, ecological and institutional changes that react to each other. During the acceleration phase, there are collective learning processes, diffusion and imbedding processes.
4. A *stabilisation* phase where the speed of the social change decreases and, while learning, a new dynamic balance is reached.

It is essential that different social processes play a role during the various phases. It is also important to realize that fundamental changes do not necessarily occur in all the domains at the same time.

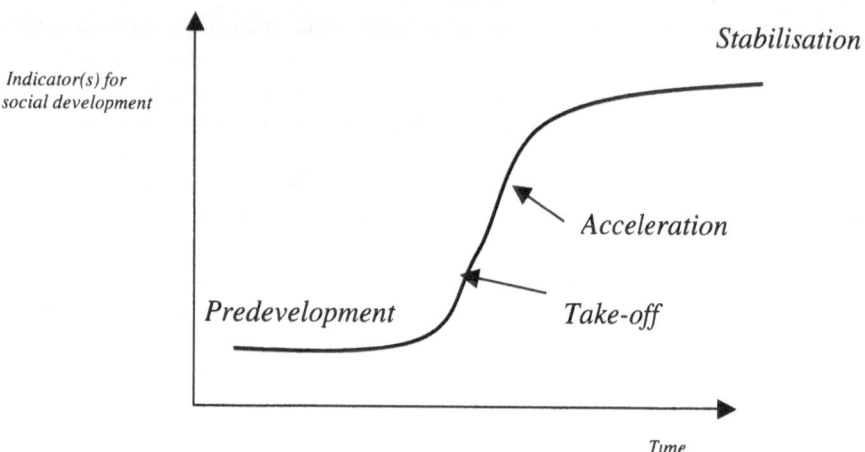

Fig. 11.2. Four phases of transition

A transition contains a period of slow and fast developments. It should be noted that the concepts of *speed* and *acceleration* are relative notions.

A transition is not a quick change in the short-term, but a gradual, continuous process. Transition processes usually cover at least a generation (25 years). Transition processes are relatively slow, because the established balance implies stability and inertia. As a result of this stability, a transition implies that an essential change of generally shared assumptions and role distribution must take place. This could be accelerated by unexpected, intermittent occurrences and events, for example, war, large accidents (Chernobyl) or an oil crisis which could speed up or slow down a transition process.

11.2
Integrated systems and evolutionary approach

If we examine the phenomenon of transition from the point of view of a system, we define a transition as a time span in which a transformation from slow dynamics to quick development and instability takes place, which finally results in relative stability again. The most important system characteristics of a transitions are i) a shift from one relative (dynamic) balance to the other; ii) the determinants of the new balance can differ from those of the previous balance; iii) the new balance is located at a different system level than the old balance; and iv) stability is a relative notion and certainly does not indicate a permanent state. Furthermore, it is a dynamic balance, i.e. there is no status quo, because a lot is changing under the surface. In general, a transition has three system dimensions: (i) the speed of change; (ii) the size of change; and (iii) the time period of change (see figure 11.3). These three dimensions determine the nature of the transition, i.e. the final balance level and the pathway to it. In principle, it is possible to have different paths to the same balance level. These paths can differ with regard to speed, size and time period. It is also possible for the same transition pattern to be realized in different ways.

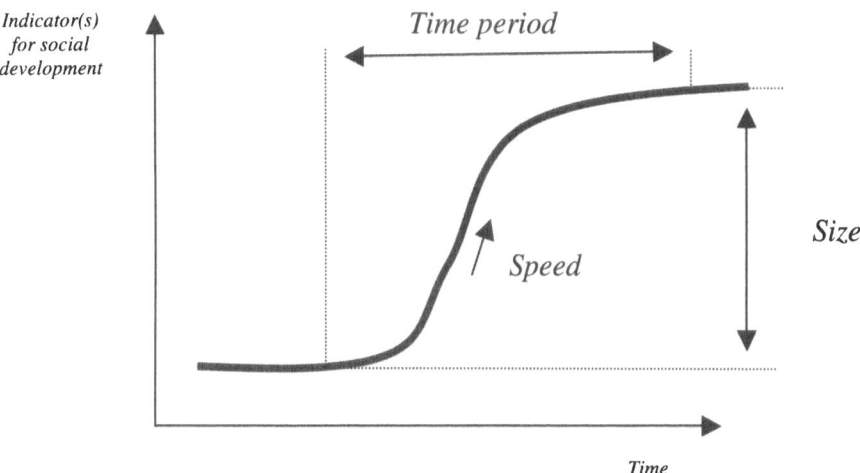

Fig. 11.3. The three system dimensions of transition: speed, size and time period

There is also strong dynamics where positive and negative feedback mechanisms can strengthen or weaken the speed of transition. Analytically, transitions are characterized by strong non-linear behaviour. During the quick period of growth, the acceleration is mainly the result of positive feedback mechanisms in the system that influence each other and which strengthen the output signal. On the other hand, negative feedback mechanisms can weaken the output signal. These are often the determinants of a new balance phase.

The system approach implies thinking in terms of *stocks* and *flows*. Stocks are aspects of a complex system that change relatively slowly over a long period of time (with regard to the total volume). Stocks are described in terms of quantity and quality. Important characteristics of stocks with regard to transition management are the amount of influence, the response time, the amount of renewability, functions and the role of actors. Flows are aspects that change relatively quickly in the short-term. Flows show the relationship between stocks. A difference can be made between material flows and information flows. GNP is an example of a flow indicator, because it measures all the short-term economic transactions. On the other hand, the total amount of economic capital goods is an example of a stock indicator. In the Netherlands, approximately NLG 100 billion per year is invested in fixed assets, of which 40 per cent is in production resources and the rest is in houses, buildings and other infrastructure. For nature and the environment, emissions to water, soil and the air can be regarded as being flow indicators. The quality of water, soil and air, as well as the size and quality of nature expressed in biodiversity, on the other hand, are examples of stock indicators, because these give an impression of the long-term state of nature and the environment. Examples of socio-cultural stocks are social cohesion, the structure of the population, lifestyle, cultural identity or the political climate. There are hardly any indicators available for these stocks. On the other hand, there are plenty of flow quantities within this domain which includes the various aspects of living, working, leisure time and health.

A transition is the result of long-term developments in stocks and short-term developments in flows. Since stocks change slowly, the dynamic pathway of a transition is characterized by a logistical curve, or an S curve. Every domain has its own dynamics. Cultures only change slowly, just like ecological systems. Economic changes, however, take place in the short-term and are usually determined by the life span of capital goods. Institutional and technological changes are somewhere in between. The whole picture, therefore, forms a hybrid mixture of fast and slow dynamics. The various time axes shift over each other and constantly influence each other. The tempo and the direction of the entire dynamics are, to a great extent, determined by the slowest processes, i. e. by the developments in stocks.

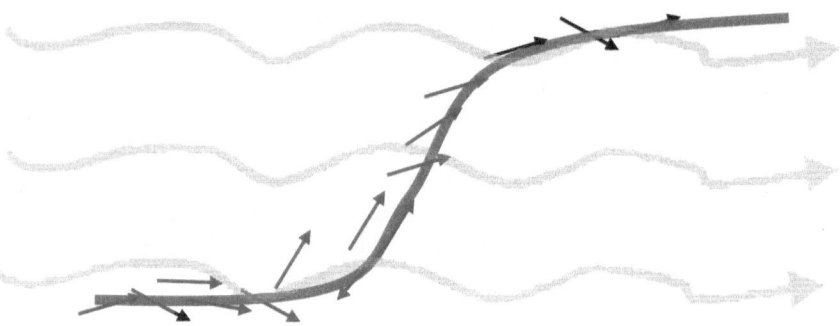

Fig. 11.4. A transition is the result of long-term developments in stocks and short-term developments in flows

The concept of transition could be used at different spatial scale and aggregation levels, such as countries or regions, companies, sectors or society as a whole. In this way, developments in the course of time, or between different countries or regions, can be followed and compared to each other. The concept of transition offers many possibilities for its use in different applications. Roughly three different levels can be distinguished in the social organisation: micro, meso and macro. The micro level contains individuals or individual actors (companies, environmental movements). The meso level contains networks, communities and organizations. Finally, the macro level contains conglomerates of institutions and organizations, e. g. a nation or an organization of states (such as the United States). This division of micro, meso and macro levels fits closely with the division which is used to describe socio-technical systems, namely the division into niches, regimes and (socio-technical) landscapes. Although the division into niches, regimes and socio-technical landscapes originates from research into and the description of technology and the social function of socio-technical systems, it also appears to be useful for the analysis of broad social changes (Geels and Kemp 2000).

The socio-technical landscape (macro level) relates to material and immaterial elements at the macro level: material infrastructure, political culture and coalitions, social values, worldviews and paradigms, the macro economy, demography and the natural environment. The second level, that of regimes (meso level), relates to the dominant practices, rules and shared assumptions (of relevant problems and solution directions). The meso level, therefore, relates to interests, assumptions and practices which can be interpreted as (implicit) rules and standards for forming the starting points for the actors' way of thinking and acting which, according to studies, appears to be geared towards system optimization. The niche level (micro level) relates to individual actors or technologies, for example. At this level, variations to and deviations from the status quo can occur, such as new techniques, a divergent form of governance or a different social practice. Alternative technologies are developed at the micro level, particularly in so-called niches. Within these niches, there are learning processes with regard to innovations, new practices or behaviour. As a result of these niches, options can be developed from ideas to alternatives. There is a process of variation and selection at this micro level resulting in path dependencies, which means that a number of other options are excluded once a certain path is taken. If the path dependency is so strong that all other possibilities are excluded, then we have a *lock-in*. The variation and selection processes are, on the one hand, dependant on the choices of individual actors, and on the other, they are partly determined by developments at the meso and macro levels. The existing regimes at the meso level often slow down the processes of change, whilst on the other hand, changes in regimes can produce breakthroughs and stimulate a transition (snowball effect). A characteristic of the macro level is that it relates to relatively slow trends and developments. Developments at the macro level can, on the one hand, play a role in speeding up or slowing down a transition, whilst on the other, changes in world views and methods of control at a macro level can produce a transition. It is as if the macro *landscape* forms gradients that channel certain paths (see figure 11.5).

From a micro perspective, this means that a number of individual actors (individuals, companies, local governments) can create *steppingstones* that make it possible for these actors to function as a catalyst for supporting the transition process. Innovations in technology, behaviour, policy and institu-

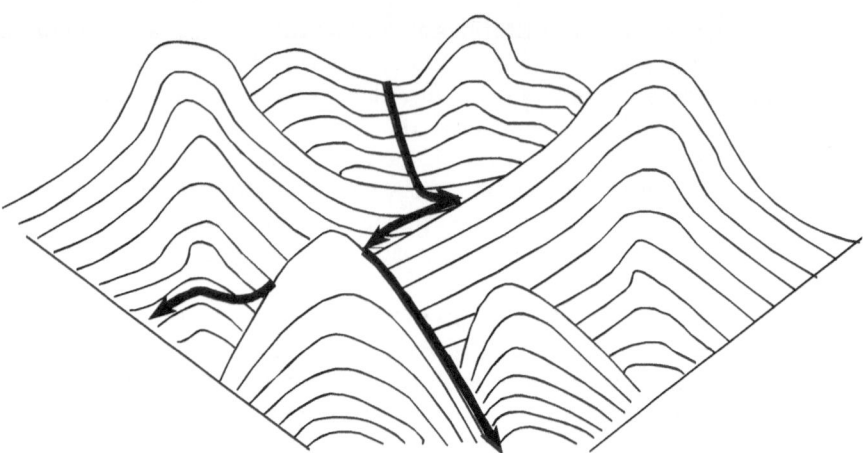

Fig. 11.5. The macro landscape channels micro and meso developments (from Sakal 1985)

tions, forms of society or the market can remain at the micro level for a very long time before they break out. Certain innovations develop at the micro level but do not break out. This is an example of invisible change in the existing social balance. If a transition is started from developments at the micro level, then a transition has the following development: it forms and stabilizes an alternative upon which both micro and meso level learning processes take place. This finally results in the creation of a support basis, so that alternatives can break out and, as a result, a take-off for a transition can be generated at the meso and macro level. On the other hand, such a take-off at a micro level can also be produced, or stimulated, through developments at the meso and macro level (for example, a change in ethics, institutional changes and changes to regimes).

Changes to regimes can occur through two different mechanisms. On the one hand, pressure from the social surroundings can lead to the discussion of an existing regime, whilst on the other, learning process concerning alternative options and the forming of new actor networks in niches can produce *bottom-up* changes to regimes. It should be noted here that existing regimes are often rather inert and embedded. In other words, regimes can slow down a potential take-off during the pre-transition stage, or even block it. Regimes, therefore, play an important role in supporting transitions from the bottom. The interaction between the different levels is shown in figure 11.6.

There is, therefore, a layered development and mutual influence underlying transitions. The multilevel perspective of transitions implies that changes only break through if the developments at one level follow on from the developments already taking place at the other levels. A transition is not only the result of developments in different domains, a transition is also the result of the interaction between developments at the micro, meso and macro level.

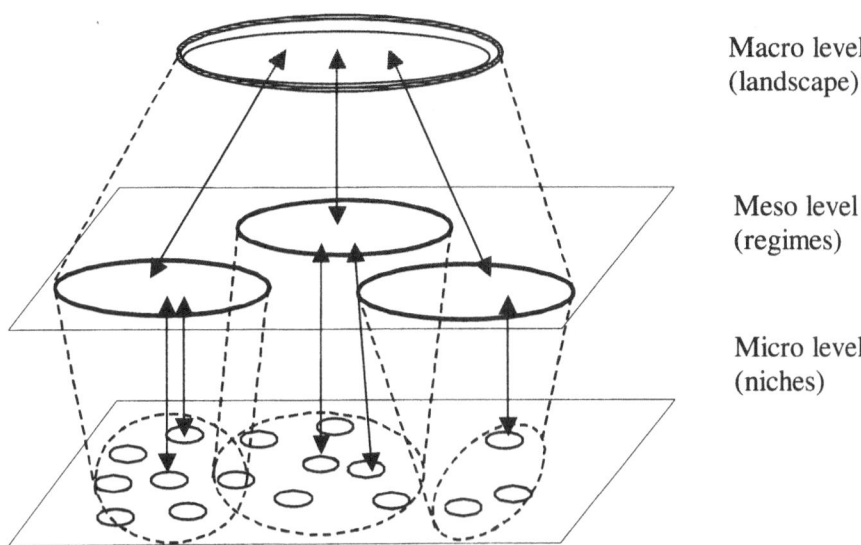

Fig. 11.6. Multilevel perspective (from Geels and Kemp 2000)

The pairing of developments at different levels gives insight into the how and why of the slowing down or speeding up of the transition process.

To summarize, we can say that a transition is a societal transformation process that has the following characteristics:

- It concerns a structural change to society (or a complex subsystem of society).
- There are large-scale technological, economic, ecological, socio-cultural and institutional developments that influence and strengthen each other.
- It is a long-term process that covers at least one generation.
- There is interaction between developments in stocks and flows.
- There are interactions between development at different scale levels (*niches-regimes-landscape*).

11.3
Example: energy transition in historical perspective

A very important transition in the Netherlands was the transformation from an energy supply that was fully based on coal to the present situation where petroleum and natural gas are the most important sources of energy. This transition has had large consequences for the extraction of raw materials, conversion technologies, the uses and even the politics and social relationships in the Netherlands. Based on a background study by Verbong (2000), we will briefly analyse the dynamic mechanisms behind this energy transition in terms of the multiple causes, and the role of the government in this transition.

What strikes first is that the transition period was short. The acceleration phase of the transition occurred quickly, in approximately 6 years. However, the preparation to this break-through, the so-called pre-development phase, was considerably longer, lasting some twenty years and took place in the nineteen twenties and thirties. In the US and in Europe petroleum was exploited and useful applications for gas, which was initially a by-product of oil extraction, were invented during that period. The rise of so-called distance gas (gas produced in large coke factories) in the nineteen twenties and thirties showed that large-scale production and distribution of gas was technically possible and economically attractive. This period can be considered as a predevelopment phase, in which the transition to an energy supply based on petroleum and natural gas was prepared. The transition itself did not take place until after the Second World War.

The form and dynamics of this transition were resulted from intertwining autonomous dynamics, developments at various scale levels and actions from a number of actors. At the macro level a number of landscape factors were important for the success of the transition. At the international level, a change in gas supply was taking place. The dominant position of coal was quickly taken over by petroleum and natural gas. Large amounts of gas were discovered, among which a large gas field in Slochteren in 1959. The price of coal experienced large fluctuations during this period. Competition from cheap coal from other countries, particularly the United States, placed constant pressure on the profitability. At the start of the nineteen sixties, coal mining became a loss making industry. In 1965, the Dutch Ministry for Economic Affairs, Den Uyl, announced the closure of the Dutch mines. Fast exploitation of the Dutch gas field in Slochteren was necessary, because the general expectation was that the price of energy would fall sharply due to the rise of nuclear energy. The rapidly increasing prosperity and the small social resistance to the technology push of natural gas provided a suitable background for the transition in the making.

An important regime factor was the public-private regime of the natural gas supply. The understanding gradually increased that the government should play a more central role in the supply of gas, in particular to provide concentration and centralization in order to stop the splitting up of the gas supply. This resulted in the establishment of a State Gas Company for the distribution of gas and a National Gas Company for the supply of gas, but the local councils and the semi-nationalized companies (Hoogovens and DSM) refused to give up their power. However, after tough negotiations with the oil companies Shell and Esso the gas supply became the monopoly of the Gasunie (Gas Association), whose shares were owned by the state and the oil companies. Local councils were incorporated. They were permitted to continue to take care of the distribution under the supervision of the Gasunie and under the conditions stated by them. Other parties were no longer tolerated. Hoogovens was bought out and DSM was included in the Gasunie on behalf of the government to simplify the closing of the coal mines.

At the niche level, it was important that a market presented itself for experimenting with natural gas, viz. the housing market for heating and cooking. The condition of the houses was poor by international standards. Houses were uncomfortable, did not have insulation and were poorly heated, so that in a technological and market respect a (large-scale) niche was present. Due to the big housing shortage, the government gave priority to quantity and not quality. The social desires for hygiene (warm water for showers/baths) and central heating (heat the whole house instead of only the living room) had already changed, but had not yet been realized.

In the nineteen sixties, the Netherlands quickly switched to natural gas for heating and cooking. This operation showed all the characteristics of a Delta plan: at a high speed and in around 6 years, all the steps that led to a transformation of the gas supply were based fully on natural gas and controlled by the Gasunie. The public was made aware of natural gas via smart advertising campaigns: besides the financial advantages, which were small for consumers because the Gasunie sold the gas at the market value, the much higher comfort and user friendliness were emphasized. To a limited extent, the smaller load placed on the environment from burning natural gas also played a role; for example for market gardeners.

What can we learn from this transition example? The transformation from coal to natural gas in the Netherlands is an example of a transition which has been guided by the Dutch government. The government had clear objectives and sub-objectives, which resulted in a very quick and relatively smooth transition, especially seen in the international context.

The government stimulated and accelerated this process by developing a type of Delta plan for the transport and distribution of natural gas, which led to the transformation of the gas supply. Furthermore, the public was bombarded with sophisticated information campaigns for natural gas, where attention was explicitly given to the advantages of natural gas (comfort, easy to use and less damage to the environment).

However, one should not forget the long pre-development phase of decades, which operated largely at the international level, and could hardly be steered by the Dutch government. This underlines also the boundaries and limits of the possibilities for the government. All landscape lights were green: the energy prices, the revolution in the international energy supply, the discovery of largely quantities of natural gas, the economic prosperity, and the small resistance against the technology push of natural gas.

These were factors that could not be directly influenced by the government but played an important role in realizing this energy transition successfully.

11.4
Transition Management

Where so far the emphasis laid on the description, understanding and explanation of social complexity via the use of the transition concept (transitional thinking), the emphasis now lies on transitional actions, or transitional management. Transition management offers a new action perspective which can be used to give form to the transitional way of thinking. The transitional action perspective is based on a different, more process orientated, philosophy of guidance where uncertainty, complexity and coherence are core concepts. Transition management is aimed at long-term anticipative and innovative thinking and acting (at least 25 years), as well as at system innovation and system improvement. Transition management is a process of constant redirecting and guidance with others, of keeping a finger on the pulse and of learning (learning-by-doing and doing-by-learning). In a conceptual sense, stocks are being directed and flows are not (or much less).

These characteristics of transition management have been derived from the properties of transitions. Transition management can be summarized in terms of the following characteristics:

- Long-term thinking as a framework of consideration for the short-term policy.
- Thinking in terms of more than one domain (*multi-domain*) and different actors (*multi-actor*) at different scale levels (*multi-level*).
- Guiding and redirecting through learning processes (*learning-by-doing and doing-by-learning*).
- Trying to bring about system innovation and system improvement.
- Keeping open a large number of options (*wide playing field*).

The aim of transition management is not so much the realization of a specific transition: it may be that it gradually appears that sufficient results are obtained via improvements and that problems are not so bad as they were thought to be, or even get solved automatically. The aim of transition management is to provide an active contribution to the design of a transition. Transition management is more about redirecting so that better insight can be gained into the complexity of a certain theme or problem and to make the relationship with other problems clearer. This leads to a different kind of solution which is much more robust in the long-term. Transition management is not orientated towards gaining control of complexity, but much more towards facilitating managing in view of complexity via an iterative process of redirecting and guidance with others.

A transition objective is necessary in order to be able to manage a transition, but this objective must be flexible. The final objective is not fixed, but is gradually created through a social process. It is important that both the transition objective (interpreted as being a *basket of objectives*, see figure 11.7) and the final images of a transition are shared by a majority of the actors taking part in the transition process. It is also not the intention for the final image and the transition objective to be fixed. Based on new insight (*learning*), the transition objective and final images should be re-adjusted and further elaborated in so-called development rounds. Both the objectives and the final images should also be socially determined, and not only technologically. The final images of a transition contain various possibilities (options) and are intended to be quality images and not quantitative objectives. Despite the fact that transition management implies flexibility with regard to the objectives, transition management is goal orientated.

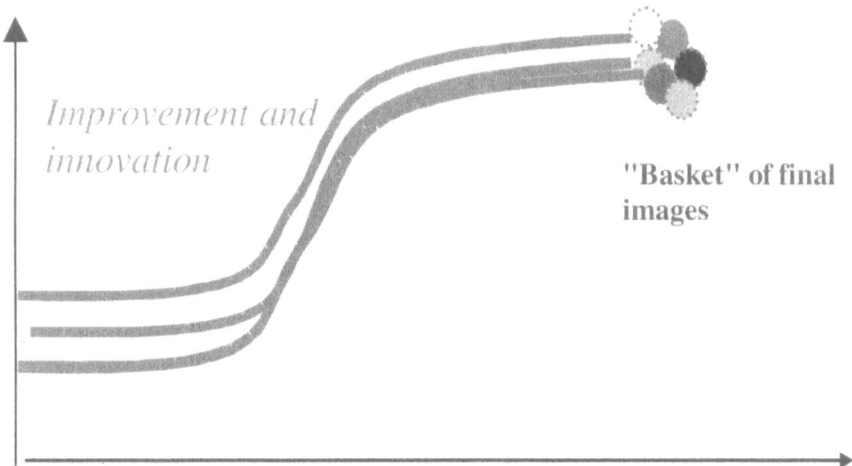

Fig. 11.7. Final objective as basket of final images

It is, subsequently, important to examine under which conditions the transition objective and the final transition images contained in this can be realized. We can investigate this by mapping out and analysing causal relationships between processes, actors, stocks, flows and events. A transition can succeed if there is diffusion, learning processes and processes of imbedding in the form of acceptance and mutual adaptation. A transition usually fails if there is destabilisation due to unfilled expectations or shifts in the environment. In this way, the transition perspective can be an aid in the integration of a policy.

The space in which the actors are able to manoeuvre must also be big enough. Within this space, the actors can then determine and take up their negotiating position. A thorough actor analysis is, therefore, a condition for the success of a transition process. The various action perspectives of the actors can then be traced and made explicit. Not only the acting perspectives of the existing, influential actors, but also those of the less important players who might become important in the future can be traced, something that might be called *niche participation*.

Important elements for transition management can be derived from this description of transition management.

- Choosing a collective transition objective.
- Exploring final images for a transition objective.
- Formulating interim objectives.
- Evaluating and learning during development rounds.
- Creating social support.

The above-mentioned steps are not taken sequentially, but form a cyclical process where the order of the steps is, to a large degree, determined by the interaction with social actors in a collective learning process of development rounds. The various transition management steps are discussed briefly below.

Transition objective

It is important to recognize that a multitude of policy aims and actor aims collectively form the transition objective. A transition objective is, therefore, multi-dimensional in the sense that it contains different objectives that are attractive to various actors and sectors and also contains more than one final image. As a result, various transition paths are possible. Formulating a transition objective is, therefore, firstly a task for society. Furthermore, it must provide guidance for the way of thinking and acting, which means that a transition objective must be concrete and tangible enough. The transition objective must leave space for choices, be evaluated and, if necessary, be re-adjusted. The transition objective sketches ambitions through the use of integral quality images.

Whereas classic policy-making tried to translate the social risks into quantitative standards and objectives, transition policy aims at more flexible, semi-quantitative or qualitative objectives. This because the traditional target-setting is doomed to fail in case of many current complex, multi-scale problems, such as climate change or sustainable development, of which the associated risks can no longer be expressed in fixed, purely quantitative objectives. Transition policy involves target-setting based on integrated risk analysis which estimates the social risks for a number of core stocks (Rotmans et al. 2000). These risk estimates concern an acceptable change in the quantity and/or quality of stocks (e. g. health, ecosystems and capital goods). A combination of these risk estimates produces the conditions under which a policy can be created. The estimates of the various types of risk are subjective, since the risks are surrounded by structural uncertainties. This uncertainty legitimizes an estimate from different perspectives (van Asselt 2000). Such an approach to transition risk is called a corridor approach (Rotmans and den Elzen 1993). A policy corridor represents a *safe* scope for policymaking, built up from policy conditions which are derived from multi-dimensional risk estimates. The policy corridor indicates the margins within which the risks may be considered as being acceptable, taking into account the uncertainties in the risk estimates. It represents a certain scope for policymaking within which social actors are able to manoeuvre.

Transition images

The transitional way of thinking is an aid for the development of a long-term version which can be used as a framework for formulating short-term objectives (such as emission reductions) and for evaluating the existing policy. Such a long-term vision provides the basis for visionary final images of the transitional pathway. The final transitional image should appeal to the interests and imagination of a broad range of social actors. Because one final image won't satisfy the ambitions of all social actors, multiple final images, in the form of a "basket" of final images, are necessary (see figure 11.7). Inspiring final images are useful (such as "putting man on the moon") for mobilizing social actors. The transition images also need to reflect the ambitions with regard to the innovation level to be achieved in the social subsystem under concern.

The final images could be adjusted as a result of what has been learned by the players from the various transition experiments set out. The participatory transition process is therefore very much a goal-seeking process, where both the transition goals and corresponding transition images change over time. So a change in transition goals and images merely reflect the learning process where the players have gone through. This in contradiction to the so-called "blueprint" thinking, which operates from a fixed notion of the final goals and the concomitant images.

Interim objectives

Figure 11.8 shows similarities and differences between current policy and transition management. In both cases, interim objectives are used. In transition-management these are derived from long-term objectives, which is not always the case for the current policy. The biggest difference, however, is in the interpretation of the concept of interim objectives and the way in which these are dealt with. In the current policy, interim objectives are often quantitative (usually quantitative emission objectives) and based on an optimal path from interim objectives to the final objective. For transition management, however, the interim objectives are compound: a combination of semi-quantitative or qualitative objectives, learning objectives and process evaluation. In other words, the interim transition objectives contain intrinsic policy objectives (which at the start can look like the current policy objectives, but later will increasingly appear to be different), process objectives (quality of the transition process, perspectives and behaviour of the actors concerned, unexpected developments) and learning objectives (what has been learned from the experiments carried out, have more options been kept open, re-adjusting options and learning objectives). In short, interim transition objectives contain three components: an intrinsic, a process and a learning component.

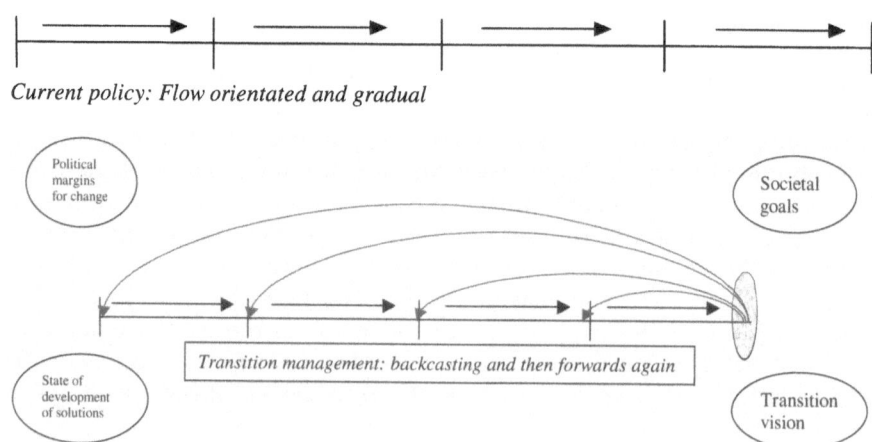

Current policy: Flow orientated and gradual

Fig. 11.8. Short-term versus long-term policy

Evaluating and Learning

Transition management uses so-called development rounds where what has been achieved in three different aspects is evaluated. The social actors who take part in the transition process evaluate in each interim round the set interim transition objectives, the transition process itself and the transition experiments.

Firstly, the set interim objectives are evaluated to see whether they have been achieved; if this is not the case, it is analysed to see why not. Have there been any unexpected social developments or external factors which were not taken into account? Or have the actors involved not complied with the agreements that were made?

The second aspect of the evaluation concerns the transition process itself. The set-up and implementation of the transition process is put under the microscope. How do the actors concerned experience the participation process? What is the communication process like and how is this maintained? Are there other actors who should be involved in the transition process? Are there other forms of participation which must be tried out?

The final issue for evaluation is the amount of learning and enrichment which has taken place in the previous period. A special point of attention is what has been learned from the experiments carried out to stimulate the transition. What has been the most important learning moments and experiences? Has this led to new knowledge and new circumstances? This last aspect, in particular, is important in development rounds. The central question in development rounds is: "What have we learned and obtained and how do we continue from there?". The idea of development rounds is, therefore, a process approach where use is made of learning moments, experiences, new knowledge and new circumstances. The new insights resulting from this can lead to new basic starting points and, consequently, to new choices. The innovating aspects of these development rounds in transition management lies in the fact that evaluation and readjustment is carried out on the basis of both content, process and learning effects.

Creating social support

Transition management means that social support is created in practice as part of a collective learning process. The explicit aim of transition management is to actively use the knowledge and expertise of social actors to create support as a result of a collective learning process. Support is created and options take shape by means of a social process of development rounds. Transition management implies the institutionalization of such an iterative and interactive social process. It is important for us to realize that transition management is not only a matter for the government. The transitional way of thinking should also be adopted by social actors as an action perspective in order to be able to trigger the necessary social process. Other processes with actors other than the usual actors and institutions are, therefore, necessary for transition management.

11.5
Transition Management in relation to the current policy

Transition management must not be seen as a frontal attack on the current policy. On the contrary, the current policy can be made to fit into a transitional policy. Transition management is supplementary. Transition policy gives added value to the existing policy by placing it in a more long-term perspective. In this way, transition management attempts to imbed the current policy in a long-term perspective that is characterized by a vision of innovation and coherence. Running or proposed policy actions are looked at from their contribution to a transition. Transition management is a proactive, long-term policy, particularly aimed at anticipative and innovated thinking. Instead of *putting out fires*, transition management is a form of guidance, which is applied to tendencies and weak signals, the so-called seeds of changes and innovation. The concept of transition places short-term policy in a much longer term perspective: i.e. the perspective of one, two or three generations

(50–100 years) instead of the time horizon of a maximum of 5–10 years, which is often used in the current policy. Transition management breaks through the old contrast between short-term and long-term policy. Transition management contains an integration of long-term thinking and short-term action because it offers a long-term framework for short-term policy.

Transition managers accept that a substantial delay occurs between the moment of guidance and the visible effect of it, i. e. there is a long response time. The current policy is particularly intended for the visible short-term effects. In the short-term, the largest visible gain is usually obtained via the improvement of existing technology, such as the idea of smoke desulphurizing units at coal burning power stations or CO_2 collection and storage. However, such an approach of scoring gains in the short-term is insufficient for complex social problems. This does not mean that transition management rejects the improvement of existing systems. It says that you must aim to achieve *both* system improvements and system innovation. Improvement and innovation do not exclude each other: cleaner cars can go hand in hand with innovative public transport systems. Transition management uses improvements as a stepping stone on the way to innovation, particularly during the pre-development and take-off phases. This implies that existing systems have to be written off, if it appears that improvement does not accommodate such a transition potential.

A characteristic of transition management is that it tries to achieve innovation without too much destructive friction (in the form of social resistance and high economic costs) by a gradual approach. The rationale behind the gradual approach of transition management is that a transition can be brought about by a gradual transformation of a system, and not by substituting an existing system. A new element can be added to an existing system, for example, in order to solve a specific problem. The introduction of this new element leads to changes and learning processes within the system. As a result of the changes, new *bottlenecks* will appear which, in the course of time, stimulate the development of new concepts, ideas, insight, methods and techniques. Furthermore, it is possible that the learning processes lead to the discovery of new possibilities. The consecutive improvements, innovations and learning processes, gradually transform the existing system so that finally a new system is created. This means that transition management should be orientated towards stimulating so-called *two-world* options during the pre-development phase and the take-off phase: options that are valuable both in the existing system and in a system that satisfies the transition objectives. Joining in with *ongoing dynamics* is often easier than forcing changes. Transition management, therefore, implies refraining from active stimulation of and (large-scale) investment in improvement options which only fit into the existing system and which, as a result, stimulate a lock-in situation.

Based on what has been previously mentioned, transition management can be described as follows:

Transition management = current policy + long-term vision + coherence + short-term action for the benefit of learning processes and keeping options open + process management (development rounds and network management)

In general, transition management tries to gradually increase the pressure on the existing system, whilst, at the same time, alternatives are explored through learning processes. Keeping open options and exploring alternatives is not the same as *not choosing*. Choosing for transition management means that we do not want time to make our choice, but transition management means taking time when time is available.

11.6
Actor analysis: the role of the government

The choices taken by social actors can stimulate, slow down or even block a transition. Therefore, the major social actors and their action perspectives have to be mapped out. Here we will in particular focus on the role of the government as actor. Which role can the government play in managing a transition? And what possibilities does the government have to stimulate a transition?

The government interacts with other social actors, but has its own responsibilities and resources. Historical research shows that it is crucially important to have a clear objective, or at least a vision of the future, to realize a transition. This is an important task for the government. The government is able to inspire and mobilize the other actors with a vision which is appealing to the imagination and which is supported by clear final images translated into objectives. Here, it is important for the government to listen carefully to the other actors in order to recognize and translate the final images and objectives which are thought to be inspiring and appealing by the other players.

Transition management, therefore, also means that the government can take the lead. Not by enforcing choices, but by inspiring a collective learning process and by encouraging other actors to think along and participate. The lower authorities, such as provincial and municipal authorities, also have a role to play in transition management. They are closer to the citizens than the national government, the local situation can permit radical experiments (such as a car-free town centre or city heating) and they have been assigned their own tasks in areas which are often relevant with regard to social transformations, such as environmental planning, house-building, the environment and waste.

It is, therefore, a misunderstanding to suppose that the government cannot play a strong role in this, although it is true that the government will fulfil a different role to the one in the past. Processes such as globalization, liberalization and privatization set the tone for the current social dynamics. On the one hand, this means that the government transfers a number of responsibilities to the market. On the other hand, the government must fulfil a much more powerful guiding role, for which it has a number of possibilities. A process such as liberalization requires a continuously changing and dynamic system of rules and laws in order to guarantee that a real market is created and not a monopoly or oligopoly. The government can also take a number of initiatives and generate stimuli to make the market more attractive to newcomers. The government creates the boundary conditions within which the market can operate, but can also create market opportunities and favourable conditions to which companies can react (also through network management). Stimulating experiments (niche management) and encouraging new technology are also some of the important options the government has at its disposal. The decision-making for the last option (new technology) will increasingly shift to the European level.

So the role of the government is different in each phase of the transition process. In the preparation phase, the emphasis lies on keeping a wide playing field, organizing and stimulating discussions with social actors and the strategic stimulation of niche-experiments, thus a role as *catalyst* and *director*. In the take-off phase, it is important to actually mobilize the actors in the direction of the collectively formulated transition objective, so *facilitating* the take-off phase. In the take-off phase and the acceleration phase the government has to stimulate learning processes and small-scale experiments. This can be achieved by forming an agenda, forming communal images concerning what is desirable and possible, creating niches or by anticipating the actor's interests. So this means a role as *stimulator*. In the stabilisation phase, the guidance is mainly orientated towards imbedding in order to prevent backlashes that exist in various member countries, so a role as *player* and *consolidator*.

So overall, the role of the government in transition management is plural: player-facilitator-catalyst-director, depending on the stage of the transition. The most effective (but least visible) is the guidance in the pre-development phase, and to a lesser extent, in the take-off phase. In the pre-development phase, it is important to keep a wide playing field and to promote variation. In the take-off phase, it is important to use the momentum well. Much more difficult is the guidance in the acceleration phase, because the direction of development in this phase is mainly determined by reactions which reinforce (or weaken) each other and which cause autonomous dynamics, so that processes become more rapid. It is then still possible to adjust the direction of development, but it is almost impossible to reverse a development direction.

Limitations to the role of the government. Although an important role is granted to the government in transition management, there are also clear boundaries and limits to the possibilities of the government. Firstly, there are a number of external factors (landscape factors), such as the development of energy prices, on which the government only has a limited influence. Furthermore, there are numerous structural and cultural limits to the possibilities of the government. The political structure in Europe places important boundary conditions, just as with the relationships, on the liberalized market.

A point that has not been elaborated in this paper, but which is important in forming policy, is that the government is not a monolithic block. It consists of various institutions and organisations which each have their own aims and interests. Finally, there are also socio-cultural factors which influence the government's possibilities to act. Social developments are increasingly the result of multi-actor processes where the government may well be an important player (actor), if not the most important, but where, in spite of this, it is definitely not able to push through developments in a unilateral and top-down manner. Creating and maintaining support for transition objectives is important in this.

To summarize, it can be said that the role of the government in transition management is very important. This role is twofold, namely concerning the contents and the process. The content role mainly concerns developing and carrying out a vision and the objectives related to this. The process role is aimed at stimulating and organizing the transition process, mobilizing the social actors concerned, creating opportunities and challenges for transition participants and creating boundary conditions within which the transition process can operate. Despite the fact that the government is a very important actor in the realization of social transitions, if not the most important, it should be aware of its limitations. External factors, social developments at the micro and macro levels, socio-cultural limitations and the international political structure, have a great deal of influence on the government's possibilities to act. Furthermore, there is no such thing as *the* government, because it consists of complex institutions and organisations. The view given above of the role of the government is in line with ideas about *modern governance* (see Kooiman 1993 and Teisman 2000), although this paper puts more emphasis on the combination of guidance according to the contents and process (instead of only guidance according to the process, see Heuvelhof en Van Twist 2000).

11.7
Case study: the transition to a low-emissions energy infrastructure

The transformation to a low-emission Dutch energy supply can be interpreted as a typical example of a societal transition. It concerns a structural change of an important social sub-system, namely the energy infrastructure for the transport and distribution of energy. This structural change of the energy infrastructure does not so much concern a system optimization, but rather a system innovation. A number of important boundary conditions will be set by the other domains which can either slow down or strengthen a transition to a low-emission energy supply. The economic domain places importance on the consideration between costs and benefits (affordability, return, green investments), the socio-cultural domain finds health, safety and reliability of delivery important, and from the ecological point of view, the risks for nature and the environment are important. On the other hand, this large-scale transformation of the energy infrastructure will in the longer run lead to social changes; it will imply institutional and social-cultural changes, as well as changes in lifestyle.

Furthermore, there is a general consensus that a transformation of the energy supply is a long-term concern, which experts estimate will require at least two generations, or 50 years. Finally, it is clear that global and European developments will have a large influence on the future Dutch energy supply. The energy issue is pre-eminently an international problem. Nevertheless, a Dutch point of view has been chosen for elaborating this case study. The above justifies an analysis of the transformation to a low-emission energy supply from the transitional perspective. Firstly, it is important to determine which phase of this transition we currently find ourselves in. It is generally considered that both in the Netherlands and internationally, we are still in the pre-development phase.

Next, the most important opportunities and obstacles for realizing the low-emission energy transition can be mapped out. An important hindrance is the large availability of fossil energy sources, leading to low energy prices and relatively limited investment in alternative sources of energy. Furthermore, the energy sector is one of the driving forces of the Dutch economy, so that substantial intervention in this sector may have large economic consequences. The current liberalization of the energy market is leading to a short-term vision from energy companies, which is mainly orientated towards cost savings. These same energy companies also fear a lock-in, in the sense that they are scared for placing all their eggs in the wrong basket (i.e. choosing the "wrong" energy technology).

Another important obstacle is that there are no clear owners of the problem. As a result, the companies and institutions causing the CO_2 emission have no real desire for change. Also with regard to the policy, there does not appear to be a direct need for a transition policy: recent studies show that the Netherlands could theoretically satisfy the Kyoto climate objective and any possible accentuation of this without a fundamental change in the energy system taking place (ECN, RIVM 1999). Finally, from a social point of view, there does not appear to be an immediate need to change the current energy supply, since there is no direct reason in the form of calamities. This ensures that there is relatively little support in society for an energy transition.

In contrast to these serious obstacles, there is a number of opportunities to get a low-emission energy supply off the ground. Firstly, it can be stated that the current energy supply is not sustainable. The most non-sustainable aspect of the current energy supply is formed by the current and future environmental effects, varying from anthropogenic climate change and biodiversity to the large-scale degradation of natural resources. This is related to the growing notion in society that, in the long-term, a change to a more sustainable energy supply is necessary. This also agrees with the recent advice from influential Dutch advisory boards who state that the energy infrastructure must change fundamentally in the long-term. An argument specific to the Netherlands is that the Netherlands, as an energy intensive country, is (too) vulnerable due to an increasing dependence on one single energy fuel, which can culminate in a technological monoculture. Therefore, a country such as the Netherlands is sensitive to political instability surrounding the energy supply, such as was demonstrated by the unrest concerning the high diesel prices. Furthermore, deferment of the energy transformation only shifts the problems to later generations, because the future possibilities for the energy supply are, to a large extent, determined by the current investment in R&D for energy technology. Finally, the Kyoto Protocol can also be seen as an opportunity and challenge to give shape to an actual energy transition. The objectives of the Kyoto Protocol to be realized could be seen as an impulse for actual system innovation. The Kyoto Protocol functions then as a catalyst for realizing the required system innovation in the energy supply.

Energy transition management

Here, we will follow the step-wise approach for transition management as formulated in section 11.4. The first step is the selection of a collective transition objective. Such a transition objective needs to be multi-dimensional and not only quantitative. From the social-cultural viewpoint safety and reliability of delivery are important requirements. The ecological risks could be specified in CO_2 emission reductions. A low-emission energy supply is often translated in terms of far-reaching CO_2 reductions, of the order of 50 per cent with regard to 1990, which is to be realized over a period of 50 to 100 years. In this study no estimation is made of the economic costs and benefits of such a low-emission energy supply.

The second step concerns exploring the final images of energy transition, which is primarily based on a recent study by the Dutch Energy Centre ECN into the future possibilities of a low CO_2 emission energy supply (ECN 2000). This has resulted in three final images for the future Dutch energy supply. Firstly, the *Status Quo* final image, where the current energy infrastructure remains intact, but the final energy fuels are made from renewable energy resources (solar, wind and biomass). In this image, it is mainly the supply side of the energy supply that changes. Methane, oil and electricity remain the final energy fuels, so that apparently not a lot changes for the final user. There are, however, a lot more conversion steps necessary, particularly for biomass and coal, where the primary energy fuels are both renewable and 'clean' fossil fuels (use of fossil fuels with storage of CO_2 in empty natural gas fields or coastal seas). The second final image is that of *The Netherlands as Hydrogen Land*, where there is a large-scale transformation of the Netherlands from a natural gas country to a hydrogen country. In this final image, hydrogen is the dominant final energy fuel, particularly for industry, transport and built-up areas. This requires a thorough adaptation of the current natural gas network, so that, for example, cars are able to run on hydrogen. The third final image is that of *The Netherlands as all electric society*. Here, the role of electricity as the final energy fuel is dominant in all sectors of society. This also requires fundamental transformation of the current energy infrastructure, including a large-scale electricity network in order to allow cars to run on electricity, for example.

These three final energy images form a reflection of a possible interpretation of the energy supply in the Netherlands. They are indicative of the applicability of certain final energy fuels. They are not disjunctive and each contains a central variant (central production of energy) and a decentralized variant (production of energy takes place close to where the energy is used). The final energy images are, however, one dimensional and of a technological nature. Real transition final images have a social dimension and are, by definition, multi-dimensional. To *convert* these final energy images into social final images, an image should be sketched of the social imbedding of such technological final images. The social, cultural, institutional and environmental contexts are extremely important in order to be able to give an image of a coherent pallet of possible social changes which is as well balanced as possible. Without such a social context, insufficient support will be gained from the social actors involved in the transition process.

The ECN analysis shows that all three final energy images can lead to a 50 per cent reduction in CO_2 emissions, but only under the motto of "all hands on deck". This means that a great deal of effort is required from all the renewable energy sources (solar, wind and biomass), but also that large-scale use should be made of clean fossil fuel energy. Furthermore, nuclear energy plays an important role in all three final images of energy and a principal role is assigned to the saving energy. This means that such a large-scale reduction in CO_2 emissions, independent of the chosen final energy image, can only be realized through a combination of factors: renewable energy, clean fossil fuel energy, nuclear energy and saving energy. What is remarkable, is the large degree of dependence on grown of imported biomass in all of the final energy images. The acreage necessary for the production of biomass amounts to approximately 1–2.5 times the total surface area of the Netherlands.

Since the costs involved with the realization of the various final images are ignored, it is difficult to make judgements about the feasibility of the various options. However, something can be said regarding the advantages and disadvantages of the various options. At first sight, the status quo final image offers a lot of advantages, since the existing infrastructure can be preserved, although an exorbitant quantity of biomass is required. Furthermore, a strong preference for a decentralized future energy supply in combination with the current infrastructure does not appear to be compatible with the important role which central energy technology is expected to play in the realization of the desired environmental objectives. The hydrogen society final image has the advantage that it can be entirely CO_2-free. Furthermore, there is considerable enthusiasm for such an advanced technological future image. On the other hand, such a fundamental changeover requires a great deal of time, sophisticated control and a great deal of effort. The electrical society final image has the advantage that a gradual transfer to low CO_2 emission, or possibly a CO_2-free energy supply, can take place in the long-term. There is, however, not a great deal of enthusiasm for this, partly as a result of the risks related to it (breakdowns, disasters) and the sidelining of a number innovative technologies currently being developed.

Formulating interim objectives is the third transition management step. This allows you to describe the various transition paths behind the final energy images. A transition management strategy can be outlined by linking the chosen final energy images to the various transition paths. If we look at the characteristics of an energy path, a couple of things catch the eye. Firstly, there is no one-to-one relationship between the transition path and the final transition image. It is possible for a low-emission energy transition to carry a number of final energy images. On the other hand, various energy transition paths can lead to the same final image. The energy transition will not be an intermittent development, but a development which will take place gradually. A large number of small steps forwards (*jolts*) over a couple of decades allow the development to take place (seemingly) unnoticed, but visibly.

To manage the low-emission energy transition, it is necessary for all the formulated final transition images (status quo, hydrogen and electricity structure) to be kept open. This means that one final image is not chosen at an early stage, but that all three images remain in the picture for a long time. Only with the passage of time (decades) will it become clear which final energy image comes across as being the most dominant (see figure 11.9). The other options then gradually disappear from the picture, although a combination also remains possible. The coming and going of options is an evolutionary process, which is partly autonomous and which can partly be influenced through anticipation and timely redirecting. Via a strategic process of system innovation, the government can influence this within a continuously changing economic, technological, environmental and institutional context.

The current policy is orientated towards observing agreements, such as the Kyoto Protocol in 2010. Both the Kyoto policy and the Kyoto[+] policy are not examples of energy transition management. The Kyoto objectives for the Netherlands can initially by obtained because the emphasis is placed on the reduction of non-CO_2 greenhouse gases and on CO_2 reduction abroad in the period 2000–2010. In this way, the national CO_2 emissions in 2010 can

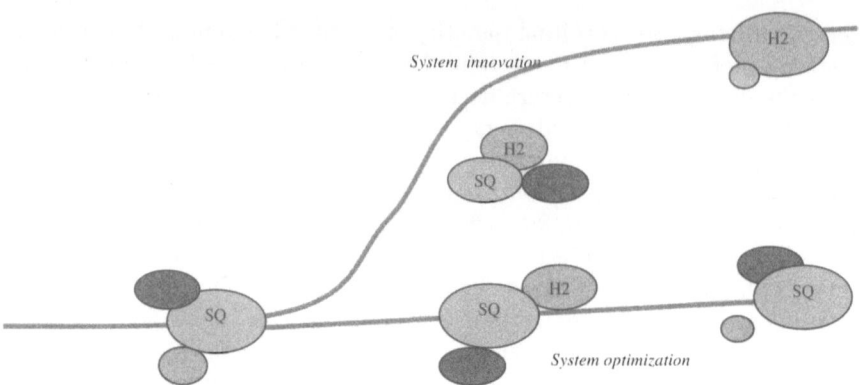

Fig. 11.9. Keeping open transition images in the course of time

still be 9 per cent higher than in 1998 (current emissions). After this, during the period 2010–2020, a CO_2 reduction of approximately 13 per cent can be realized via a great deal of effort, so that the final CO_2 emissions in 2020 can be approximately 6 per cent lower than in 1998 (ECN, 2000). A CO_2 reduction of 6 per cent compared to now appears to be obtainable without a fundamental revolution in the energy infrastructure taking place. On the other hand, an enormous effort by the policy is required for this, with a high use of renewable energy, a great deal of energy saving at a high cost.

But with a focus on the medium term, reaching no further than 2020, the current energy infrastructure (based on oil, gas and electricity) remains unchanged. This means that the time to change after 2020 is reduced from 50 to 30 years. By doing so, two of the three transition images are removed from the picture: the hydrogen and the electricity societies, causing a so-called lockout. In other words, the current Kyoto policy is orientated towards realizing concrete short-term emission objectives via a process of system improvements and not by system innovation. A continuation of the current policy (Kyoto+) is orientated towards the medium-term emission objectives, also via system improvements. Such a policy does not require fundamental changes to the energy infrastructure and steers a course directly to the status quo final image. This implies a lockout of the other final energy images, which, as a result, automatically disappear from the picture (see figure 11.10).

Continuation of the current climate policy in the long-term almost automatically leads to the exclusion of promising innovative energy options. On the other hand, a specific energy transition image does not have to be in any way contrary to the current policy. A specific energy transition image assumes the Kyoto policy for the short-term, but places it in a long-term perspective (50 years). In this way, the Kyoto policy can be seen as a step towards long-term system innovation. In the short-term, nothing is turned upside down and no forced change to the energy infrastructure takes place.

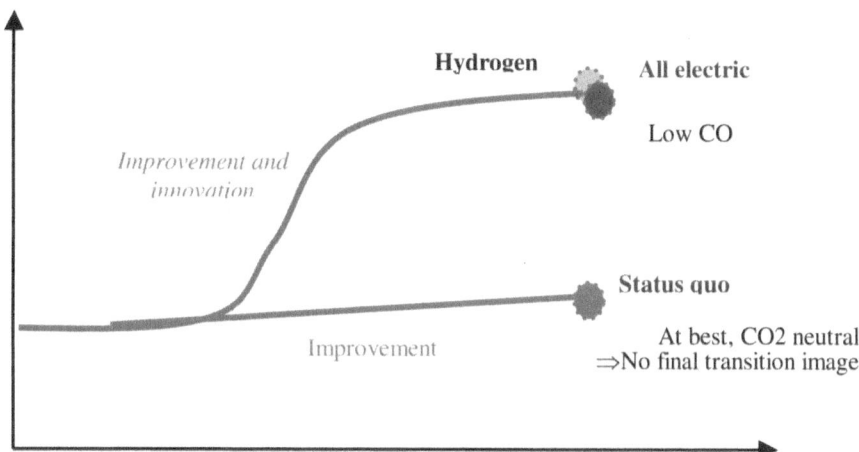

Fig. 11.10. Kyoto process and the process of system innovation

The fourth step in managing the transition involves the evaluating and learning process in development rounds. This requires a permanent and innovative participatory process with involvement of current and future social actors. In this way the government could organize a broad energy transition debate by taking the lead in mobilization of the social actors. The challenge is to maintain this long-term participatory process in a manner that it remains attractive to all social actors engaged. The government also could promote a discussion about the energy transition at an international level (EU and UN).

An important step involves setting up an innovative package that consists of innovative research and experiments at the local and regional level with new energy technologies, new institutions and new instruments. Contrary to the current innovation package as part of existing climate policy with technological and instrumental proposals, such an innovation package as part of energy transition management, however, is also aimed at proposals for institutional innovation (long-term consultation structures, the policy with regard to knowledge institutions and the policy with regard to research programs), stimulating learning process and the coherence between the innovation elements. Current and new research could, for example, be grouped by the final energy images which they can contribute to. Furthermore, it may be useful to set up complementary projects to research the non-technical barriers for a low-emission energy transition. The entire technological and instrumental packages can also be extended through action which keeps the playing field wide and innovative research can be set up by systematically carrying out experiments. Examples may be the specific investment in PV, not as a solution, but as a mode for keeping a wide playing field: case studies into the positioning of solar PV in medium sized Dutch towns, case studies into the problems with the location of large solar PV stations. Other interesting learning experiments involve research into the large-scale positioning of wind turbines in the North Sea, which the Netherlands, Belgium, Great Britain, Denmark and Norway are all involved in. Options which are being used within the framework of the current energy policy, such as stoking conventional power stations with biomass, could be used to investigate the effects of large-scale biomass use and cultivation more empirically: how much space is needed for this, what are the consequences for biodiversity, what are the consequences in terms of the balance of power in the world.

It is also important to identify *two-world* options in order to gain insight into the possible transition paths: how can we get from the existing system via different steps to a new energy supply? Could the current natural gas network be made suitable for more hydrogen? Could CO_2 storage play the role of interim phase in a large-scale transformation process? How would it be possible to stimulate the use of equipment which uses an alternative energy fuel in order to stimulate innovation from the side of demand?

Possible criteria for evaluating short-term actions and experiments from the point of view of transition are:

- Do the action and experiments taken provide insight into the coherence between the socio-cultural, economic, nature and the environment, and institutional dimensions of a transition?
- Are the possibilities for innovation and transition paths explored through the action and experiments taken?
- Do the action and experiments taken contain potential for learning? If so, which relevant learning objectives can be recognized with regard to transition management?
- Are other actors stimulated to become joint owners of a problem and to adopt the transition objective as their own action perspective?

The final transition management step is aimed at creating social support. This is a continuous process that runs through the entire course of transition process as a recurring theme. The most important players in the current energy field of influence should be mapped out for this by performing an actor analysis. In the Dutch energy field the following social actors can be distinguished: energy companies, the government, users (consumers) and the social organisations (NGOs). The background image is one of a strongly liberalized and privatized energy market, where both the market situation and the role of the players are liable to change. This gives an unsettled playing field which is characterized by a large number of uncertainties upon which social actors must operate.

Against this background, energy companies take up a position of wait and see. As a result of the large amount of competition, they address themselves to the short-term and wish to make maximum use of their investments in the current energy infrastructure. Energy companies consider 5 to 10 years to be long-term and there is a definite fear of a *lock-in* in the sense that energy companies are scared of placing everything on the wrong energy fuel. At the same time, the understanding is growing that it is becoming increasingly important to develop a specific technology strategy.

The government is still searching for its new role in a quickly changing market. On the one hand, the government can create the boundary conditions within which the national energy market can operate, as well as create market opportunities which companies can play along with. On the other hand, the government can also set limits to the movement of the national energy market, particularly by imposing environmental obligations. With regard to the influence of the government on the future energy infrastructure, a distinction must be made between the production and the distribution of energy. Where the involvement with the production of energy could be small, the government will be very clearly involved with the distribution of energy: a high-quality, safe and relatively cheap distribution network is extremely important for the consumers. From the social sense of responsibility that the government has always shown with regard to the energy supply, undesired developments in the field of distribution can be redirected or corrected. This, therefore, appears to be primarily a duty of the government and not so much a task for the other actors in the energy playing field. While the European context becomes increasingly more important, the role of the local councils also becomes increasingly important, particularly through new experiments with renewable energy (particularly wind and biomass, but also solar boilers).

The energy users (consumers) will get a free choice in choosing their energy supplier. The future will show to what extent the users are sensitive to the strategy of customer relations which many energy companies say they wish to introduce. The users have a fairly strong position in the continuing development of decentralized energy technology. They can stimulate new developments (for example, solar boilers, PV), but they can also slow down new developments or even stop them (for example, heat pumps). Social organisations have an indirect, but important, influence on the future energy supply. They can function as a catalyst in a sustainable energy market which is to be created. Their influence could increase in a liberalized market, because they are close to the consumers who can choose their own energy supplier.

11.8
Conclusions

In this paper we have explored the notion of transition and applied it to the case of the future energy supply in the Netherlands. A major conclusion of our study is that there appears to be a vacuum where it is not clear at all who will provide guidance for realizing a transition to an innovative, emissions-low energy supply. All social actors point to the government as being the potential leader: the government largely determines the direction in which the energy infrastructure shall move. In the field of renewable energy, the government has an extensive set of instruments, ranging from subsidies and convenants about energy efficiency and certificates for green electricity. However, the government's policy has hardly been developed in other energy fields, such as the transport sector, the chemical industry and the heat supply. So, in spite of the limitations the national government has to face, the government plays a key role in advancing a low-emission energy transition. The government largely determines the momentum and the direction of the energy transition. The guiding role of the government is twofold: it is both concerned with the process and the contents. An energy transition policy contains the current climate policy, but adds something essential to it: a long-term vision, an innovation impulse by carrying out specific experiments, coherence between the short-term and long-term policy and between micro and macro developments.

However, our analysis shows that it won't be easy to realize such an energy transition. Apart from the macro-scale factors which have to be favourable, it also requires a double role of the government. In process terms the government has to facilitate the transition process, whereas in terms of contents, the government has to inspire the other social actors, as director and catalyst. The guidance for the process of a transition will require a different form of participation, however, with participation from other actors. Via a process of so-called *niche participation*, new players who are as yet insignificant but who may become important in the future can also become involved in the process. The government can also push a smaller party forwards as the leader and even have a more controlling function itself. Furthermore, the government also plays an intrinsic role where it can exercise a visionary function. Such an intrinsic function is necessary to develop a consistent long-term vision which is shared by all the parties concerned.

The research reported in this paper indicates that the transition concept can be used to structure complex societal dynamics, in such a way that it indicates levers for action. It was hypothesized that the perspective of transition management can be used as new approach to the management of complex policy problems. This is only a first step. In order to underpin the concept of transition management more research, especially through case-studies and using advanced theorizing on modern governance, is needed. With this paper we hope to invite researchers from different disciplines to engage in this challenging endeavour.

Acknowledgements

This paper is the result of collaborative research on transition and transition management between the International Centre for Integrative Studies (ICIS) and the Maastricht Economic Research Institute on Technology (MERIT) that took off with an interdisciplinary advisory project for Dutch Ministries (esp. VROM, EZ, V&W and LNV) in the context of the 4th Nature and Environment Plan (NMP[4]). We would like to thank Frank Geels (Twente University), Geert Verbong (Technical University Eindhoven) and Kirsten Molendijk (ICIS), who substantially contributed to the project report "Transities en transitie-management". We furthermore would like to thank the energy experts who participated in an expert workshop on the energy transition, i. e. Geert Verbong (Technical University Eindhoven), Wim Turkenburg (Utrecht University), Tom Kram (ECN) en Ron Wit (CE-Delft). Last but not least, we would like to thank the NMP[4] working group 'Transitie naar een emissie-arme energievoorziening', i. e. Silvie Warmerdam, Paul van Slobbe, Peter Aubert, Bart Thorborg, Ruben van der Laan, Alfred van Hoorn, Frans Vlieg, Frans Vollenbroek, Paul Tops and Job van den Berg. Our intensive and inspiring high-level discussions were essential in shaping the concept of transition-management. We would furthermore like to thank the NMP[4] project team and especially projectleader Cees Moons for his continuous trust and support. Finally we would like to thank the participants to the various seminars and lectures (August 2000–December 2000) on transition-management at various ministries; their questions helped us to further crystallise our ideas.

References

General Energy Commission Algemene Energieraad (1999) Overheidsbeleid voor de lange termijn energievoorziening (Translation: Government policy for the long-term Energy Supply). Advice to the Dutch Ministry of Economic Affairs, July 1999

Davis K (1945) The world demographic transition. Annals of the American Academy of Political and Social Science 237 (4), pp 1–11

Dutch Ministry for Housing and Regional Development and the Environment (VROM) (1999) Uitvoeringsnota Klimaatbeleid. Deel I: Binnenlandse maatregelen (Translation: Policy memorandum of climate policy. Part I: National regulations). The Hague, The Netherlands

ECN (1998) Nationale Energie Verkenningen 1995–2020 (Translation: National Energy Explorations 1995–2020). ECN-C-97-081, March 1998

ECN (2000) Energietechnologie in het spanningsveld tussen klimaatbeleid en liberalisering, Energie Centrum Nederland (Translation: Energy technology in the field of tension between climate policy and liberalization, Dutch Energy Centre). ECN-C-00-020, May 2000

Geels F, Kemp R (2000) Transities vanuit socio-technisch perspectief (Translation: Transitions from a socio-technical perspective), background document for chapter 1 of Rotmans et al (2000)

Heuvelhof EF ten, Twist MJW van (2000) Nieuwe markten en de rol van de overheid (Translation: New markets and the role of the government). ESB dossier, Liberalization of network sectors

Kemp R, Schot J, Hoogma R (1998) Regime Shifts to Sustainability through Processes of Niche Formation. The Approach of Strategic Niche Management, Technology Analysis and Strategic Management, 10(2): pp 175–195

Kemp R, Truffer B, Harms S (1998) Strategic Niche Management for Sustainable Mobility. In: Rennings K, Hohmeier O, Ottinger RL (eds) (2000), Social Costs and Sustainable *Mobility – Strategies and Experiences in Europe and the United States*. Physica Verlag (Springer), Heidelberg, New York, pp 167–187

Kemp R, Rip A, Schot J (2000) Constructing Transition Paths through the Management of Niches. Forthcoming in: Garud R, Karnoe P (eds) Path Creation and Dependence. Lawrence Erlbaum Associates Publ

Kooiman J (ed) (1993) Modern Governance. New Government-Society Interactions. Sage, London

Ness GD, Drake WD, Brechin SR (ed) (1993) Population-Environment Dynamics: Ideas and Observations. The University of Michigan Press, VS

NMP⁴ (2000) Wokken op Waterstof: Transitie naar een emissiearme energievoorziening (Translation: Cooking with hydrogen: Transition to a low emission energy supply). Final report of the NMP⁴ workgroup 'Duurzame economie: het duurzaam voorzien in de behoefte aan energie en mobiliteit' (Translation: Sustainable economy: durably meeting the need for energy and mobility), 23 May 2000, The Hague, The Netherlands

Notestein FW (1945) Population, the long view. In: Schultz TW (ed) Food for the world. Chicago, pp 36–57

Rip A, Kemp R (1996) Towards a Theory of Socio-Technical Change, mimeo UT, report prepared for Batelle Pacific Northwest Laboratories, Washington, D. C. An edited version has been published as book chapter, 'Technological Change'. In: Rayner S, Malone EL (1998) Human Choice and Climate Change. An International Assessment, Vol. 2, Batelle Press, Washington D. C., pp 327–399

Rotmans J, den Elzen M (1993) Halting Global Warming: Should Fossil Fuels by phased out? In: Lal M (ed) (1993) Global Warming. Concern for tomorrow. Tata Mc Graw-Hill Publishing Company Limited, New Delhi

Rotmans J (1994) Transitions on the move, Global Dynamics and Sustainable Development. Dutch National Institue of Public Health and the Environment (RIVM), Bilthoven, The Netherlands

Rotmans J et al (1995) TARGETS in Transition, Global Dynamics and Sustainable Development. Dutch National Institute of Public Health and the Environment (RIVM), August 1995. Bilthoven, The Netherlands

Rotmans J, Kemp R, van Asselt MBA, Geels F, Verbong G, Molendijk K (2000), Transitions and Transition Management: the case of a low-emission energy supply. ICIS-Report, Maastricht, October 2000

Sahal D (1985) Technological guideposts and innovations avenues. Research Policy Vol 14, pp 61-82

Teisman G (2000) Sturen als ontwikkelingsopdracht (Translation: Guidance as a development assignment). In: Report of discussions with scientists during the NMP⁴ process. RMNO, The Hague, The Netherlands, pp 64–67

van Asselt MBA (2000) Perspectives on Uncertainty and Risk: the PRIMA approach to decision support, Kluwer Academic Publishers, Dordrecht, The Netherlands

Verbong G (2000) De Nederlandse overheid en energietransities: een historisch perspectief (Translation: The Dutch government and energy transitions: a historical perspective), background document of Rotmans et al. (2000)

VN (1997) Critical Trends: Global Change and Sustainable Development, Department for Policy Coordination and Sustainable Development, United States, New York

VROM (2000) Transities naar Duurzaamheid: Milieubeleid als Transitiemanagement (Translation: Transitions to sustainability: Environmental policy as transition management). Vision document for Knowledge and Technology workgroup, draft 10 April 2000

VROM-raad (1998) Transitie naar een koolstofarme energiehuishouding (Translation: Transition to a low-carbon energy economy. Advice for the Climate Policy Implementation Memorandum, Advice 010, 23 December 1998

Weaver P, Jansen L, Grootveld G, Vergragt P (2000) Sustainable Technology Development. Greenleaf Publishing, Sheffield

12 The Inclusion of Stakeholder Perspectives in Integrated Assessment of Climate Change

Jeroen van der Sluijs, Penny Kloprogge

12.1
Introduction

Managing the risks of anthropogenic climate change is a societal process which has to deal with a long term complex issue under conditions of high and partly irreducible uncertainties and multiple value orientations of the many national and international stakeholders. Over the past decades Integrated Assessment (IA) has emerged as an approach to link knowledge and action in a way that is suitable to accommodate the uncertainties, complexities and value diversities of global environmental risks.

IA can be defined as an interdisciplinary process of combining, interpreting and communicating knowledge from diverse scientific disciplines in such a way that the whole set of cause-effect interactions of a problem can be evaluated from a synoptic perspective with two characteristics: (i) it should have added value compared to single disciplinary assessment; and (ii) it should provide useful information to decision makers (Rotmans and Dowlatabadi, 1997)[1].

In the early stages of its development, IA of climate change was mainly the domain of experts and technical computer models largely dominated the field. In the nineteen nineties participatory methods, in which perspectives of stakeholders of the problem are included, came to be generally applied tools in the field of IA.

Going into the nature of the problem of climate change we will argue why it is desireable and important to include stakeholder perspectives in IA. We will also show that the need to include these perspectives arose in the ongoing international policy process, as a reaction to limitations that were met by the classic expert-driven scientotechnical approach to IA. After discussing why

---

[1] Note that in practice, IA addresses only part of 'the whole set' of cause effect interactions because analysts always make a selection and simplify and because the knowledge on and understanding of causal relations in complex environmental problems is partial and incomplete.

inclusion of stakeholder perspectives is desireable, we will turn to the question how stakeholder perspectives have been included in IA-practice.

12.2
Motives for including stakeholder perspectives

Climate change is a problem in which facts are uncertain, values are in dispute, stakes are high and decisions are urgent. It is what Funtowicz en Ravetz call a 'post-normal problem'. (Funtowicz and Ravetz 1992; 1993; Ravetz and Funtowicz 1999, see also Funtowicz' contribution in this volume). Problems with these characteristics can not be handled by the 'normal', truth seeking science, but require a different problem coping strategy known as post-normal science. As long as scientific uncertainties and decision stakes are high, the aim of finding the 'resolution to the scientific puzzle' is in principle unachievable and undesirable. The questions put on science by the policy process are urgent so we cannot wait until science has the ultimate answers. Instead, post-normal science aims at common commitments to approaches for dealing with uncertainty and value diversity in complex policy issues.

Because of the many uncertainties, traditional science is not able to sufficiently legitimise drastic steps to combat climate change. The traditional dominance of "hard facts" over "soft values" has been inverted: hard value commitments have to be made, based on soft facts (Funtowicz and Ravetz 1993). The assessment of risks and the setting of policy should therefore encompass public agreement and participation. In addition, participation is desirable to enhance the quality of the scientific input in the process. For, when facing the uncertainties involved in post-normal problems, scientists are lay people as well (Funtowicz and Ravetz 1993). Stakeholders' reasoning, observation and imagination is not bounded by scientific rationality, which can be beneficial when tackling ill-structured complex problems. Consequently, the perspectives of the stakeholders can bring in valuable new views on the problem and relevant information on that problem. The latter is known as "extended facts" and includes anecdotes, informal surveys, knowledge on local conditions which may help determine which data are strong and relevant, and personal observations of environmental change which may lead to new foci for empirical research addressing dimensions of the problem which were previously overlooked. Scientific and technical discourse in post-normal problems is no longer restricted to expert communities but needs to include nonspecialist perspectives of stakeholders on the problem at hand. They play an important role in quality control of the scientific process. Making use of this valuable reservoir of scrutiny-potential requires the establishment of an extended peer community (Funtowicz and Ravetz 1996) and the inclusion of stakeholder perspectives not only in the phase where solutions are debated, but also in the assessment process that precedes it.

Reasoning from the concept of post normal science, inclusion of perspectives of stakeholders in IA is necessary for handling uncertainty, disagreement and dissent in risk assessment and the setting of policy, for the enhancement of the quality of the assessment and for legitimisation of action taken in the policy process and of the science that underpins the choice for that action.

Not only a theoretical contemplation on the nature of the climate problem leads to the conclusion that the inclusion of stakeholder perspectives in IA is essential. This need can be seen arising in practice as an inevitable response to the changing characteristics and requirements of the different successive phases of the international climate policy process. The reaction of the field of IA to this arising need for the inclusion of stakeholder perspectives is clearly reflected in the shift of IA with a purely scientotechnical analytical toolbox to IA with a toolbox with complementary analytical and participatory methods and in a parallel increase in the involvement of social sciences in climate risk assessment.

The development of international climate policy can roughly be divided into six periods (adapted from Sprinz and Luterbacher 1996):

- the foundational period, during which scientific concern about global warming developed;
- the agenda-setting phase, from 1985–1988, when climate change was transformed from a scientific into a policy issue;
- a pre-negotiation period from 1988 to 1990, when governments became heavily involved in the process;
- the formal intergovernmental negotiations phase, leading to the adoption of the United Nations Framework Convention on Climate Change (FCCC) in May 1992;
- a post-agreement phase focusing on the elaboration and implementation of the FCCC and the initiation of negotiations on additional commitments;
- with the Kyoto protocol a new phase has started, which – in the light of Jan Rotmans contribution on transition management in this volume – may eventually be understood as a pre-transition phase (transition towards a low carbon economy).

IA started to play a role in the agenda setting phase: the 1985 Villach international Conference on the Assessment of the role of Carbon Dioxide and of Other Greenhouse Gases in Climate Variations and Associated Impacts was a milestone in bringing together experts from a multitude of scientific disciplines to arrive at an integrated view on the climate problem. Three years later, the World Conference on the Changing Atmosphere: Implications for Global Security held in Toronto marked the beginning of high-level political debate on the risks of anthropogenic climate-change. The Villach and Toronto conferences were successful attempts by scientists to put the issue of climate change on the policy agenda and to initiate the first steps of an international climate policy regime. In the pre-negotiation phase, climate IA models started to be developed (for a comprehensive historic review see: Van der Sluijs 1997).

In this phase, scenario analyses were needed to support the thinking about the question by how much greenhouse gas emissions needed to be reduced in order to manage the risk. In response to the developments in the post-agreement phase of the international climate policy regime, in which quantified emission limitations and reduction objectives were negotiated, IA model-results from different modelling groups were fed into the negotiations. Key examples of IA-results produced in this phase have been 'safe landing corridors' (tolerable windows), defined as the allowable lower and higher bounds of greenhouse gas emission scenarios, related to a set of criteria for climate policy whose purpose is to protect both the environment and the economy from disruption. With the Kyoto protocol, a long-term effort has started to design possible institutions and mechanisms whose implications are being gradually explored by international environmental diplomacy.

This phase is essential to build up the know-how and the trust relations between the parties involved which are required to develop environmental policies at a global scale. However, policies agreed upon by the international policy arena are clearly insufficient if the problem of climate change is to be effectively addressed. If effective climate policy is to emerge, actions taking place at the level of international environmental diplomacy must be combined with actions involving various kinds of stakeholders. They range from farmers to forest managers, from tourist operators to inhabitants of coastal zones, and from financial investors to ordinary citizens. Involving the latter will be necessary because climate mitigation measures will require consumer and worker co-operation as well as citizen consent to be successfully implemented.

For IA this implied a growing need for the integration of social science research, and in particular of participatory techniques into research on global change to include stakeholder perspectives in the assessment process. Social science is needed to provide knowledge about stakeholders and their ways of opinion formation, and also to provide opportunities for including the knowledge of stakeholders, their perspectives and their judgements about controversial issues in policy making.

12.3
Methods to include stakeholder perspectives

In the following sections we will further explore the question *how* stakeholder perspectives have been included in different stages of IA processes. For this purpose we will use insights and experiences from four European Participatory Integrated Assessment (PIA) projects, all addressing the climate change problem, namely, COOL, Risk Approaches, TARGETS and ULYSSES. A short description of these projects is given in boxes 1 to 4.

We will focus on the different approaches to include stakeholder perspectives that can be identified in each of these projects. More specifically, we will address for each of the PIA projects the following questions:

- Whose perspectives have been involved in the assessment?
- In what phase of the process have their perspectives been involved?
- How have stakeholder perspectives been included?
- Who has effectively been allowed to contribute what relevant wisdom in what phase of the risk management process? (boundary work)

12.3.1
Whose perspectives have been involved?

The PIA-projects discussed in this paper employ different terminology regarding stakeholders and use different (implicit) definitions of stakeholders. Some projects talk about citizen participation (e. g. ULYSSES), other projects seem not to include citizens in their demarcation of the concept of stakeholders (e. g. COOL).

Box 1 COOL

COOL is the acronym for "Climate OptiOns for the Long-term" (Berk et al., 1999), a participatory Integrated Assessment project aiming at supporting the development of long-term climate policy in the Netherlands in a European and global context. The project is subdivided into three parts: a dialogue at the national level, one at the European and one at the global level. Here, we will only go into detail on the dialogue at the national level.

In the National Dialogue dialogue workshops are formed by non-government stakeholders from major sectors of the Dutch economy. Together they develop long term strategic visions using an interactive back casting technique: starting point is a reduction of emissions of greenhouse gases of 50–80 per cent by the year 2050. The sector groups look backwards to identify path ways and obstacles to reach this goal. The participants are handed two images of hypothetical futures with different perceptions on issues as the availability of fossil energy sources and the development of economy. Experts provide the participants with state of the art knowledge and check the groups' ideas on feasibility. Website: http://www.wau.nl/cool

---

Box 2 Risk Approaches

The Risk Approaches project (Van der Sluijs et al. 2000; Hisschemoller et al. 2000) was set up with the goal to evaluate existing climate risk assessment studies to identify areas for improvement in climate risk assessment to better match the information needs of the users (policy and society) of these studies. For this purpose the project applied a special type of non-steering open interviews: repertory grid analysis (Dunn and Ginsberg 1986; Dunn, Pavlak and Roberts 1988; Dunn 2000), to elicit from stakeholders the constructs they use to think about climate risks. These constructs have in turn been used to evaluate existing risk assessment studies to identify mismatch in the extent to which they resonate with the cognitive frameworks of the stakeholders regarding the climate problem.

The repertory grid analysis enables the bottom-up elicitation of subjectively meaningful constructs employed to interpret the risks of anthropogenic climate change – that is, it permits observations of thought acting on the external environments characterized as a problem situation – in this case anthropogenic climate change. Grid-methodology avoids asking specific questions to the interviewees. The resulting constructs can be seen as dimensions that need to be addressed in integrated assessment in order to resonate with the cognitive frameworks employed by the stakeholder population. Website: http://www.chem.uu.nl/nws

Box 3 TARGETS

TARGETS: Tool to Assess Regional and Global Environmental and Health Targets for Sustainability (Rotmans and De Vries, 1997) was built at RIVM in the Netherlands to evaluate the consequences of several types of human influences simultaneously. TARGETS consists of a interlinked set of sub-models simulating population and health, energy, economy, biophysics, land/soil, and water.

TARGETS deals with value diversity and uncertainty by allowing multiple model routes depending on a typology of so-called perspectives, which guide choices about fundamental uncertainties. Three perspectives are implemented in the model, that differ with regard to key model assumptions that govern the dynamics of the natural and the social system in the model: Individualist (Nature is robust), Egalitarian (Nature is fragile), Hierarchist (Nature is robust within certain limits, but fragile beyond those limits). The perspectives also correspond to preferred management styles for policy interventions accounted for in the model (Egalitarian: prevention, Individualist: adaptation or Hierarchist: control). The user can choose to run the model in one perspective only (utopia) or to mix perspectives (dystopia). These mixed perspectives address conflicting perspectives. That is: Imagine the world would work according to worldview X, however be managed by management style Y. The authors of the model favour none of the perspectives, but use them as a means to address subjectivity in IA modelling.

Box 4 ULYSSES

ULYSSES – short for Urban LifestYles, SuStainability, and Integrated Environmental ASsessment – (commissioned by EC DG XII, part of the Fourth Framework program) has developed a procedure for citizen participation in Integrated Assessment (IA). The procedure comprises a discursive process based on the focus group method. The citizens debate climate policy and sustainable development, and they have been given access to state-of-the-art Integrated Assessment Models (IAMs) to support their debates.

Between 1996 and 1999 ULYSSES has conducted and analysed group discussions with over 400 citizens in eight European cities. Each focus group consists of 6–8 citizens and meets for five individual sessions or for two consecutive days. In the 1st session environmental problems and climate change are discussed generally and in some groups the participants are encouraged to produce collages to illustrate their concerns. In the 2nd session global issues are addressed using an IAM to stimulate the discussion. The 3d and 4th sessions focus on regional and local issues, supported by a regional IAM. In the 5th session the participants produce a "citizens report" based upon the discussions in all sessions. A group moderator guides the discussions and a model moderator introduces and demonstrates the IAMs.

Website: http://zit1.zit.tu-darmstadt.de/ulysses

For the purpose of this paper we use a broad definition of stakeholders which we adapted from the definition of the World Bank (1996): stakeholders are those actors who are directly or indirectly affected by an issue and who could affect the outcome of a decision making process regarding that issue or are affected by it. Each of the PIA projects has motivated their selections of stakeholders on the basis of the specific purposes of each of the projects. These purposes are elaborated upon in box 1–4. The projects distinguish different stakeholder groups including citizens, different sectors of economy, business, environmental NGOs, research managers, and government officials at ministries.

One of the projects, TARGETS, applies an ideal typical representation of stakeholder perspectives rather than involving actual stakeholders. Three perspectives are distinguished within the TARGETS model, that differ with regard to key assumptions about nature and society: Individualist (Nature is robust), Egalitarian (Nature is fragile), Hierarchist (Nature is robust within certain limits, but fragile beyond those limits). These perspectives have been developed as ideal types in the so-called cultural theory of risks (Douglas and Wildavsky 1982). The TARGETS group assumed that these three perspectives span up the policy relevant part of the spectrum of value diversity in the stakeholder community.

A summary of our findings regarding who's perspectives have been involved is presented in table 12.1.

Table 12.1. Stakeholder perspectives involved in each project

Project	Whose perspectives were involved?
COOL	Stakeholders from four major sectors of the Dutch economy: Built environment; Industry; Agriculture and Nutrition; Traffic and Transport
Risk Approaches	27 (potential) users of climate risk assessment in the Netherlands: business 10; environmental NGOs 5, research managers 5, Ministries 7.
TARGETS	3 Ideal typical perspectives, which are assumed to capture the policy relevant part of the societal spectrum of value-orientations: Hierarchist, Egalitarian and Individualist.
ULYSSES	400 Citizens in 8 European cities; plus a smaller sample of decision makers from public policy and the private sector, representatives from the financial industry interested in ecological investment, and media representatives.

12.3.2
In what phase of the process have stakeholder perspectives been involved?

Inclusion of stakeholder perspectives in IA can take place in different stages of the process. Figure 12.1 sketches the different phases that can be distinguished as well as the place of IA in the environmental risk management process. For our analysis here we use a subdivision in risk management functions (Kates et al. 1985) that was used by the project Social Learning in the management of Global Environmental Risks (The Social Learning Group 2001; Toth and Hizsnyik 1998) which distinguishes: risk assessment, options assessment, goal and strategy formulation, implementation, monitoring, and evaluation. Risk assessment addresses the understanding about the nature, causes, consequences, likelihood and timing of climate change. Option assessment addresses the feasibility, costs and benefits of possible options to manage a risk. Goal and strategy formulation involve the setting of management goals, the design of a package of options appropriate for achieving them, and the selection of modes for implementing those options. Implementation involves actions actually taken by various actors (including governments) to manage climate change. Monitoring includes documenting actual changes in aspects of the environment affected by climate change, and the results of management strategies and specific implementation measures. Evaluation encompasses self-conscious efforts of actors to reflect upon and evaluate their own and others' performance on climate risk management. We have added to this set of functions the function of problem framing which includes the setting of problem boundaries and problem definition. The IA process generally comprises the functions problem framing, risk assessment, options assessment, and generally assists in goal setting and strategy formulation (fig. 12.1).

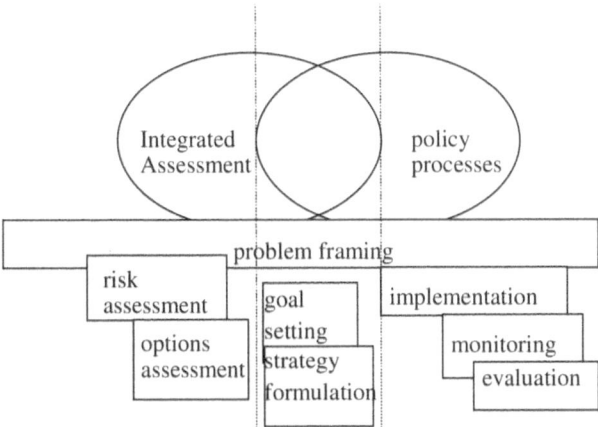

Fig. 12.1. The different phases in the process of environmental risk management and the place of IA within this broader context.

Using this framework we have mapped in what risk management functions the stakeholder perspectives were involved in each of the PIA projects we looked at. The results are summarized in table 12.2.

Table 12.2. This table maps for each of the PIA projects in what risk management functions the stakeholder perspectives were involved. Dark grey shading indicates primary involvement, light grey shading indicates some or indirect involvement.

	COOL	Risk Approaches	TARGETS	ULYSSES
Problem framing				
Risk assessment				
Options assessment				
Goal setting				
Strategy formulation				

12.3.3
How have stakeholder perspectives been included?

In the four PIA processes discussed here, a variety of methods to include stakeholder perspectives can be identified. ULYSSES applied open-ended discussions in moderated focus groups of citizens. ULYSSES has demonstrated that IA-Focus Groups can increase the possibilities for citizens to articulate their view on environmental issues in a reflective manner. This can help to bridge gaps between decisions by policy makers and citizens' views. Exploring alternative explanatory and moral frames in such processes can complement expert assessments on complex environmental issues. Focus group procedures involving open-ended discussions add a richer analysis to the traditional quantitative survey methods. In IA-Focus Groups, the discussions can broaden into directions impossible for the researchers to imagine beforehand. This is vital if we are to learn more about how the diversity of everyday life and experience affects the understanding, values, and objectives that different social groups bring to global change issues.

In the COOL project, dialogue workshops were used in which the discussion was significantly less open ended: the policy goal (80 per cent CO_2 reduction by the year 2050) was established in advance by the scientists and the policy makers. The stakeholders involved in the COOL project were not consulted when establishing this goal for the back casting exercises. Instead the participatory process focussed on the development of strategic long term visions with regard to economical and political choices needed to achieve such an ambitious transition to an economy with a low carbon intensity. This approach has the advantage that the participants are committed to engage in a highly focussed creative process, exploring concrete and feasible pathways, which could make a long-term transition to a low carbon economy attainable. Another advantage is that the process utilises the most salient expertise of the participants, namely local knowledge from the sectors of economy in which the desired transition ultimately has to be implemented.

The TARGETS project applied a top-down (from Cultural Theory) ideal typical approach to include in the assessment the societal spectrum of value-orientations of the stakeholder community. The coupling of cultural theory to model assumptions in TARGETS has been path-breaking in the sense that it acknowledges the possibility of multiple problem structures – both at the level of causal mechanisms behind the problem and on the level of preferences and values – and in that it recognizes the legitimacy of different perspectives on the science. On the other hand this method has a shortcoming in that it restricts the problem structuring to three different static problem definitions in terms of pre-defined ideal-typical categories. By doing so, TARGETS hampers the integration of differing perspectives and differing value positions into new ways of looking at the problem.

TARGETS was also used actively in focus groups in the ULYSSES project. These experiences have shown that the approach chosen in TARGETS to convey uncertainty was often misunderstood or rejected by citizens. The underlying subjectivism was seldomly appreciated as an attempt to promote an honest and pluralistic debate, but rather as an unwillingness of scientists to take sides and stick to (unpopular) positions (Dahinden et al. 1999).

Where the TARGETS project adopted a top down approach, the Risk Approaches project applied a bottom up approach to capture stakeholder perspectives. The Risk Approaches project has focussed on the problem framing and risk assessment phases. The rationale behind this project has been to minimize occurrence of what Dunn calls type-III error: assessing the wrong problem by incorrectly accepting the false meta-hypothesis that there is no difference between the boundaries of a problem, as defined by the analyst, and the actual boundaries of the problem (Raifa 1968, redefined by Dunn 1997). The remedy proposed by Dunn to cope with this type of uncertainty is context validation (Dunn 1998, 2000). Context validity refers to the validity of inferences that we have estimated the proximal range of rival hypotheses. Context validation can be performed by a participatory bottom-up process to elicit from stakeholders rival hypotheses on causal relations underlying a problem and rival problem definitions.

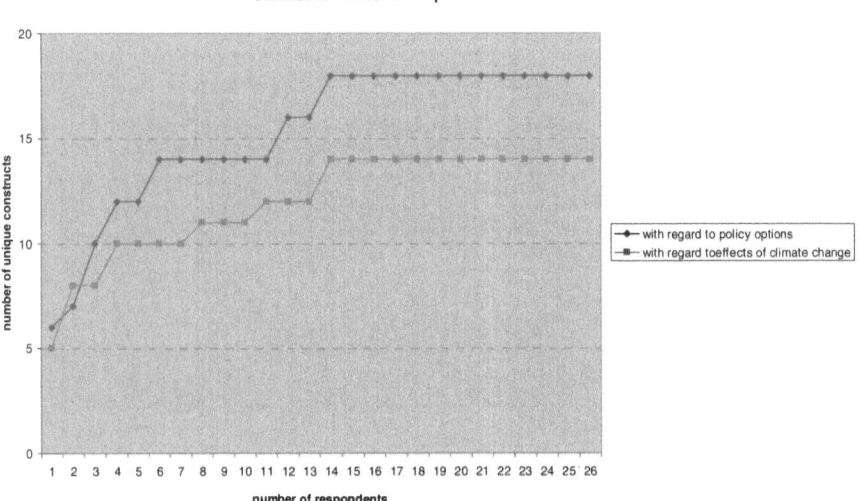

Fig. 12.2. The cumulative number of unique constructs with regard to policy options (upper line) and effects of climate change (lower line) elicited from stakeholders interviewed in the Risk Approaches project clearly demonstrates the saturation rule.

One could argue that such an open bottom up inclusion of stakeholder perspectives (be it rival hypotheses on a problem or rival constructs used to judge a problem) is endless as there are as many different perspectives as there are different citizens, and one does not know when one has covered the true range of perspectives in the stakeholder community. However, it has been shown that in projects where such elicitation processes were used, the cumulative distribution of unique rival hypothesis flattens out after a limited number of stakeholders, usually somewhere between 20 and 25.

In the Risk Approaches project this so called 'saturation rule' was also observed (figure 12.2). This confirms the validity of the assumption that inclusion of a sample of about 20–25 stakeholders (chosen in such a way that they reflect the broad range of stakeholder groups) is sufficient to approach the proximal range of constructs of the stakeholder community represented by the sample.

Table 12.3 summarizes the various approaches for inclusion of stakeholder perspectives used in the PIA processes discussed and maps on what key features these methods differ from each other.

Table 12.3. Differences in approaches to the inclusion of stakeholder perspectives in Integrated Assessment projects.

	Core-method	Stakeholder involvement	Bottom up or top down	Openness to participation per phase of [*]		
				PF	RA	OA
COOL	Backcasting form pre-established long term goal in dialogue workshops	Active	Bottom up / Top down	Closed	Closed	Open
Risk Approaches	Repertory grid analysis	Passive	Bottom up	Open		
TARGETS	Cultural theory	Passive	Top down	Closed	Closed	Closed
ULYSSES	Moderated IA-focus groups	Active	Bottom up	Open	Open	Open

[*] PF=Problem Framing, RA= Risk Assessment, OA= Options Assessment

12.3.4
Who has effectively been allowed to contribute what relevant wisdom in what phase of the risk management process?

With the increased involvement of non-scientists in the debates on climate change, boundaries between domains of competence became contested. The processes in the science/policy/society interfaces by which parts of a debate are depoliticised by defining them as belonging to the scientific domain or by which parts are politicised by defining them as belonging to the policy domain, is what the sociologist Thomas Gieryn (1983) calls boundary work. In her study on the role of scientific advisers in American regulatory politics, Jasanoff (1990) has further explored how scientists use a variety of boundary-defining strategies to establish who is in and who is out of relevant peer groups and networks of prestige or authority. Jasanoff argues that boundary work by scientists grows out of a premise that seems diametrically opposed to the concept of negotiation and yet is equally essential to the closure of controversy. By drawing boundaries between science and policy, scientists post "keep out" signs to prevent non-scientists from challenging or reinterpreting claims labelled as "science". The creation of such boundaries seems crucial to the political acceptability of expert advice. She also found that the experts themselves seem at times painfully aware that what they are doing is not "science" in an ordinary sense, but a hybrid activity that combines elements of scientific evidence and reasoning with large doses of social and political judgement. The boundaries between the domains of competence involved in a debate usually remain subject to continuous renegotiations. Boundary work results in a demarcation of domains defining who is seen as competent in specific areas of the problem, c. q. who is effectively allowed to contribute what relevant wisdom in what phase of the participatory risk management process. It is something that needs to be dealt with in participatory IA practice.

The PIA projects we looked at differ in the roles and competencies implicitly or explicitly attributed to each of the participants in the process (science, policy makers, stake holders). A comprehensive analysis of the phenomenon of boundary work is beyond the scope of this paper and requires empirical research. We will limit ourselves here to a few diagnostic illustrative observations of obvious differences in PIA project design, which reflect boundary work.

The COOL project shows the strongest boundary work leading to the sharpest demarcation of roles and competencies of participants: scientists and policy makers had set a prior goal for the backcasting excersises (namely 50 to 80 per cent CO_2 emission reduction by the year 2050). This was not a goal with any legal status, it was just a hypothetical imagined goal based on what policy makers and climate experts in the Netherlands believe to be necessary for the long term. The stakeholders role was narrowed down to discussing strategic combinations of technological options (which constitutes yet an other restriction of domains by excluding behavioural options and cultural change) to arrive at that goal. By joining the dialogue, stakeholders had to take that (hypothetical) long term goal for granted as a shared starting point. Within COOL, scientists play an active role (and they have been implicitly attributed exclusive competence) in bringing in expertise on present and future emission reduction technologies (potentials, expected costs etc.) and performing consistency checks on the scenarios that have been developed in the dialogues. Stakeholders bring in sector specific knowledge on the feasibility and implementability of technological changes.

A different approach to demarcating the domains where participant can bring in their views and wisdom can be identified in the ULYSSES project, which has been the most open projects with no boundary setting in advance. Within the ULYSSES project, the scientists role was made deliberately passive. The focus groups were moderated by a group moderator whereas the use of IA computer models in the focus group was moderated by a model moderator. The group moderator was in charge of guiding the process of focus group discussions (avoiding an expert role), while the model moderator presented the computer model and guided the specific discussions during the computer interaction period. The most important task of the model moderator is to present the computer model in a supportive way for the IA participatory debate. Participants' comments and their discussions indicate to the model moderator whether participants receive – not necessarily accept – the insights being conveyed. The role of the model moderator is that of acting as a two-way bridge between the computer model and the participants, together with providing some thematic focus, and guiding the discussion.

ULYSSES explicitly aimed at opening up the science: whatever the range of the computer model options (with regard to input and output variables, action and policy options), the model moderator was explicitly instructed not to limit discussions to what the model considers nor to the model reasoning. In this regard, the model moderator asked questions such as: "Which other aspects not included in this model do you think are worth considering?" (Dahinden et al. 1999). In that sense ULYSSES has been a typical post normal science exercise.

12.4
Discussion and conclusions

In this paper we explored motives and methods for including stakeholder perspectives in IA of climate change. We argued why it is important to include these perspectives taking in mind the post-normal nature of the climate problem: facts are uncertain, values are in dispute, stakes are high and decisions are urgent. Taking the views of stakeholders into account in the IA process is necessary to (1) better handle uncertainty, disagreement and dissent in risk assessment and the setting of policy; (2) enhance the quality of the assessment; and (3) legitimize action taken in the policy process and the science that underpins the choice for that action. We showed that the need to include stakeholder perspectives in IA can also be observed in the historic development of the role of IA. The initially successful scientotechnical approach to IA met its limitations in supporting the ongoing policy process. More and more stakeholders became involved in the debate and in the current phase of the policy process, know-how has to be acquired, trust relations have to be build between parties involved, and there has to be active support from people that are affected by policies that are considered. Here the previously mentioned motive of legitimisation can clearly be seen.

Participatory IA (PIA) processes are complex, consist of several phases and each PIA-process has its own dynamics. This makes comparison of the handling of stakeholder perspectives difficult and it makes it even more difficult to draw conclusions on what works best. In this paper we discussed four recent PIA-processes from which we identify different methods and procedures used to include stakeholder perspectives in the IA-process. Our findings allow us to make some remarks on the handling of stakeholder perspectives in IA of climate change.

In the four cases studied, we found the following techniques to include the stakeholders' perspectives: focus groups with open ended discussions (ULYSSES), dialogue-workshops using a back casting technique (COOL), repertory grid analysis to elicit from stakeholder constructs they apply to think about climate change (Risk Approaches) and the use of ideal types of stakeholder' perspectives based upon Cultural Theory of risk (TARGETS).

One of the major differences is whether the stakeholders are involved actively in the IA process, i. e. taking active part in the assessment itself or passively: the perspectives of the stakeholders are used by the analysts to improve IA practices (as we have seen in the Risk Approaches project) or IA models (for instance the TARGETS model). Ideally, stakeholder perspectives are incorporated in the IA method (either analytical or participatory methods), after which stakeholders are involved in the actual IA process. The use of the TARGETS model in the ULYSSES process is an example of this. However, in this particular case, the participants of the ULYSSES focus groups using this model, were confused by the way in which TARGETS deals with uncertainty and value-diversity.

Other differences that can be seen between the cases are the use of a bottom-up (ULYSSES and Risk Approaches) or top-down (TARGETS) approach and, closey related, the openness of different phases of the IA process for stakeholder participation. The latter varies from completely open, as in the ULYSSES and Risk Approaches project, to strictly demarcated sub-domains where participants can bring in their expertise, depending on the presumed competencies of scientists, policy makers and stakeholders (as in COOL and TARGETS).

12.5
Acknowledgements

We thank James Risbey for making useful comments to the manuscript, which helped us to improve this paper. This paper has benefited from the authors participation in the ULYSSES project, supported by the European Commission, DG XII, RTD Programme 'Environment and Climate', area 'Human Dimensions of Environmental Change' (Contract No. ENV4-CT96-0212) and in the Risk Approaches project, supported by the Dutch National Research Program on Global Air Pollution and Climate Change (contract no. 954266).

References

Berk MM, Hordijk L, Hisschemoller M, Kok MTJ, Liefferink D, Swart RJ, Tuinstra W (1999) Climate Options for the Long term (COOL) Interim phase report. Dutch National Research Progràmme on Global Air Pollution and Climate Change, Report no. 410 2000 028

Dahinden U, Querol C, Jäger J, Nilsson M (1999) Using computer models in participatory integrated assessment, ULYSSES Working Paper 99–2, Technical University of Darmstadt (http://www.zit.tu-darmstadt.de/ulysses/docmain.htm)

Douglas M, Wildavsky A (1982) Risk and Culture. University of California Press, Berkeley

Dunn WN (1997) Cognitive Impairment and Social Problem Solving: Some Tests for Type III Errors in Policy Analysis. Pittsburgh, Graduate School of Public and International Affairs, University of Pittsburgh

Dunn WN (1998) Pragmatic Eliminative Induction: Proximal Range and Context Validation in Applied Social Experimentation. GSPIA working paper 001, Graduate School of Public and International Affairs, University of Pittsburgh (http://www.pitt.edu/ wpseries)

Dunn WN (2000) Using the Method of Context Validation to Mitigate Type III errors in Environmental Policy Analysis. GSPIA working paper 016, Graduate School of Public and International Affairs, University of Pittsburgh (http://www.pitt.edu/ wpseries)

Dunn WN, Ginsberg A (1986) A sociocognitive approach to organizational analysis. Human Relations, Vol 39/11, pp 955–975

Dunn WN, Pavlak TJ, Roberts G (1988) Cognitive Performance Appraisal: Mapping Managers' Category Structures Using the Grid Technique, Personnel Management, Vol. 16/3, pp 16–19

Funtowicz SO, Ravetz JR (1992) Three Types of Risk Assessment and the Emergence of Post-Normal Science. In: Krimsky S, Golding D (eds) Social Theories of Risk. Greenwood: Westport CT, pp 251–273

Funtowicz SO, Ravetz JR (1993) Science for the Post-Normal Age. Futures September 1993, pp 739–755

Funtowicz SO, Ravetz J (1996) Risk Management, post-normal science, and extended peer communities. In: Hood C and Jones DKC (1996) Accident and Design, Contemporary debates in Risk Management. UCL Press, pp 172–182

Gieryn TF (1983) Boundary-work and the Demarcation of Science from Non-Science: Strains and Interests in Professional Ideologies of Scientists. American Sociological Review 48, pp 781–795

Hisschemoller M, Boer, J de, Breukels M, Dunn WN, Sluijs JP van der (2000) Climate risk assessment: evaluation of approaches. Working document 1, Depertment of Science Technology and Society, Utrecht University

Jasanoff S (1990) The Fifth Branch, Scientific Advisers as Policy Makers. Harvard University Press, Harvard

Kates RW, Hohenemser C, Kasperson JX (eds) (1985) Perilous Progress: Managing the Hazards of Technology. Westview Press, Boulder

Raifa H (1968) Decision Analysis. Addison-Wesley, Reading MA

Ravetz J, Funtowicz SO (1999) Post-Normal Science-an insight now maturing. Futures 31, pp 641–646

Rotmans J, Dowlatabadi H (1997) Integrated Assessment Modelling. In: Rayner S, Malone EL (eds) Human Choice and Climate Change. Vol. 3. The Tools for Policy Analysis. Battle Press, Columbus, OH, pp 291–377

Rotmans J, De Vries B (1997) Perspectives on Global Change, The TARGETS Approach. Cambridge University Press, Cambridge MA

The Social Learning Group (2001, in press). Learning to Manage Global Environmental Risks: A Comparative History of Social Responses to Climate Change, Ozone Depletion and Acid Rain. MIT Press

Sprinz D, Luterbacher U (1996) International Relations and Global Climate Change. Potsdam Institute for Climate Impact Research. Report No. 21

Toth FL, Hizsnyik E (1998) Integrated environmental assessment methods: Evolution and applications. Environmental Modeling and Assessment 3, pp 193–207

Van der Sluijs JP (1997) Anchoring amid uncertainty; On the management of uncertainties in risk assessment of anthropogenic climate change Ph. D. Thesis. Utrecht University, Utrecht

Van der Sluijs JP, Lourens P (2000) Climate risk assessment: evaluation of approaches. Working document, 2, Department of Science Technology, Utrecht University, 2000

World Bank (1996) The World Bank Participation Sourcebook, The World Bank, Washington. pp. 6, 13, 126

Further volumes of the series *Wissenschaftsethik und Technikfolgenbeurteilung*: